The Book of Eccles

A Portrait of Sir John Eccles.
Australian Nobel Laureate and Scientist.

1903 – 1997

by

Mary R. Mennis
MA, MSocSc, DipEd

For the Centenary year of his Birth, 2003

Mary Mennis, born Mary Eccles, is the third daughter of Sir John Eccles. Mary has a great interest in history, with her first master's degree in oral history, and her second in anthropology/material culture, based on oral sources. Mary is the author of a number of books based on Papua New Guinean oral history.

In writing this book, Mary Mennis utilised many sources, including existing publications, interviews, letters and other old documents, acknowledgment to which is given in the text.

By the same author:

They came to Matupit.
Time of the Tauber.
Kain, Friend of Maclay.
Sailing for Survival.
Hullo Eccles
Tolai Myths of Origin. (with Janssen and Skinner)
Hagen Saga.
Archipelago of the Contented People.
History of St. Dympna's School and Parish.

Publishing history:

First edition, January 2003.

ISBN 0-9750346-0-X

Published by:

Lalong Enterprises,
(ABN 32 758 899 823)
11 Jethro St.,
Aspley,
Queensland, 4034.
Phone +61 7 3263 6327
Fax: +61 7 3263 5121.
Email: mennisb@ozemail.com.au.

Cover Photograph:

A proof sheet of portraits taken of Sir John Eccles in the early 1940's, with the text overprinted on a diagram of neurones and their synapses taken from Ramón y Cajal, 1911.

Lalong Enterprises
(ABN 32 758 899 823)

Photograph by courtesy of News Ltd.

John Eccles:

Like Odysseus, I have travelled the oceans carrying my own equipment with me like a snail with his house on his back. At each "port of call" I have had to set up and develop research rooms. But I have been fortunate to discover expert technical and engineering assistance at each of my five ports of call after Oxford; otherwise my scientific life would have faltered.

Quote from *My Scientific Odyssey.*

From family records.

Author's Note:

This book is written primarily for people who might like to know about the life of my father, Sir John Eccles, particularly students in schools and colleges. It is also written for his descendants and relations, of whom there are many in Australia and overseas so that, through these photographs, cartoons and information, they may have a record of his life for now and for their descendants. Parts of this book appeared in *Hullo Eccles,* published in 2000, which gave a much broader picture of the Eccles family in Australia from 1850 to 2000.

In July 2002, I was on my way from London to Bristol and decided to make a detour to Oxford to see the old colleges of Exeter and Magdalen. As my coach pulled out of the Victoria Station in London, a man just managed to jump on and took the seat beside me. We got talking and he told me his name was Sergei and that he was a research scientist in the Physiology Department in Oxford. When I asked him if he had heard of my father, John Eccles, who had studied in Oxford in the 1930's, he was quite excited and said that his portrait hung in the entrance hall of the Physiology laboratory and also in the Sherrington Room. He insisted I go to the Department and gave me detailed instructions on how to get there. That day, I met Professor Colin Blakemore who kindly showed me around the new Department and the entrance to the old laboratory where my father and Sir Charles Sherrington once worked. Since then, the librarian at Oxford University, Sophie Wilcox, has sent me copies of letters Sherrington wrote to my father as well as a photograph of them which appears in Chapter 2. I thank her for her help.

While Professor Blakemore visited Brisbane a few weeks later, he obtained a copy of *Hullo Eccles* and, on his return to Oxford, placed this in the Physiology library. I subsequently heard from one of my father's former students, Marianne Fillenz, who wrote from Oxford. "Last week in the Physiology department library I discovered your book on the Eccles family. I started reading it and could not put it down. It is not only a wonderfully vivid picture of your father and his life but also of his family about whom he never spoke to me. It was wonderful to complete the picture."

I hope this Portrait of my father continues to fill the gaps for some people. As I am no neurophysiologist, I must refer people who want further information on his scientific life to the work of Professors David Curtis[1] and Per Andersen[2] in their biographical paper for the *Historical Records of Australian Science* for the Academy of Science, Canberra.

Dr Curtis would agree that Sir John did his best scientific work at the John Curtin School of Medical Research in Canberra during the period 1952 to 1966. Between these years, he was knighted and received the Nobel Prize. Many years later in 1990, was made a Companion in the Order of Australia. Although he left Australia in 1967, never to return, he kept in touch with his colleagues and with various family members many of whom visited him over the years. I visited him three times in Switzerland where he spent the last twenty-two years of his life. While there in 1980, he regaled me with stories of his childhood, which I recorded, and he spoke also of his scientific successes.

In putting this story together, there are many people I would like to thank. Dr David Curtis and my older sister, Dr Rosamond Mason who both scrutinised an early version of the manuscript and made critical comments which were very helpful. My cousin, Fr Ian Howells S.J., also edited a later draft of the manuscript and gave welcome advice. Others who commented were Cameron and Joanna Hoy, Greg and Clare Mennis, Sr Kath Muirhead, Mona La Reux and Neil Howard.

Other family members helped with photographs and information including Alice Ley, Dr Peter Eccles, Judith Ryan, Frances Standing, Richard Eccles and Jim Eccles.

I would like to thank the following for permission to use the various photographs and cartoons found in this book.

- Peter Jeffrey, of the John Curtin School of Medical Research, for the photographs of Sir John from his Canberra era used in Chapters 5 and 6 and the photograph of the Eccles children with the portrait of Sir John in Chapter 8.

---

[1] David R. Curtis AC, FRS, FAA, Emeritus Professor of the Australian National University.
[2] Per Andersen, Professor of Neurophysiology, University of Oslo.

- The Editor, *The Australian*, for the photographs of Sir John in his obituary 6 May 1997.
- The Editor, Herald and Weekly Times Ltd, for the Frith cartoon, *The Rescue*, in Chapter 6.
- The Editor, *Warrnambool Standard*, for the photograph by Damian White of the unveiling of the plaque for the centenary celebrations at Koroit Primary School in the Epilogue.
- Mary Fielding for the photograph, taken by her husband, of the plaque unveiling at Warrnambool College in November 2002 in the Epilogue.
- Fran Short of the Medical School, Otago University, Dunedin, for the photographs on the page preceding Chapter 4 and for arranging the necessary copyright releases.

The schools, which my father attended, Koroit Primary School, Warrnambool High School and Melbourne High School have all been interested in this project and have given me access to school records. Alan Gregory, the Historian of Melbourne High in particular has given much time and encouragement, as has Mary Fielding, Deputy Principal of the Warrnambool College (formerly the Warrnambool High School).

My husband, Brian, spent many weeks preparing the manuscript for printing and adding the photographs to the text. Thank you Brian.

Mary Mennis (nee Eccles).
January 2003

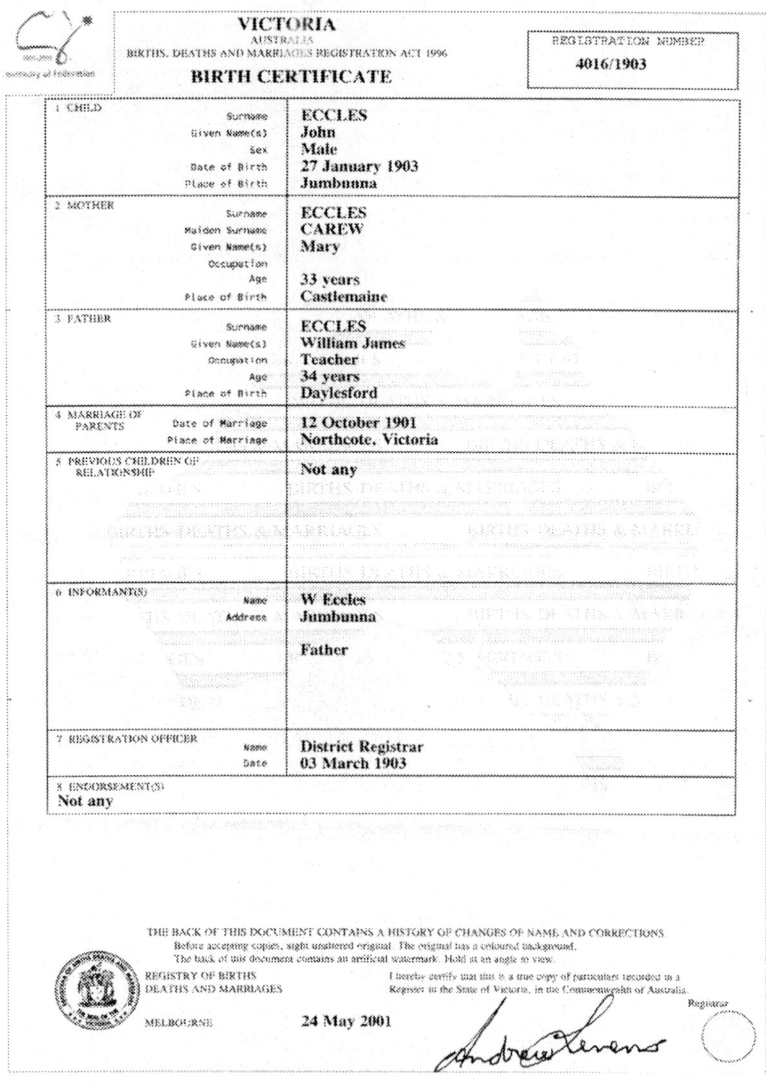

VICTORIA
AUSTRALIA
BIRTHS, DEATHS AND MARRIAGES REGISTRATION ACT 1996
BIRTH CERTIFICATE

REGISTRATION NUMBER 4016/1903

1 CHILD	Surname	ECCLES
	Given Name(s)	John
	Sex	Male
	Date of Birth	27 January 1903
	Place of Birth	Jumbunna
2 MOTHER	Surname	ECCLES
	Maiden Surname	CAREW
	Given Name(s)	Mary
	Occupation	
	Age	33 years
	Place of Birth	Castlemaine
3 FATHER	Surname	ECCLES
	Given Name(s)	William James
	Occupation	Teacher
	Age	34 years
	Place of Birth	Daylesford
4 MARRIAGE OF PARENTS	Date of Marriage	12 October 1901
	Place of Marriage	Northcote, Victoria
5 PREVIOUS CHILDREN OF RELATIONSHIP		Not any
6 INFORMANT(S)	Name	W Eccles
	Address	Jumbunna
		Father
7 REGISTRATION OFFICER	Name	District Registrar
	Date	03 March 1903
8 ENDORSEMENT(S)		Not any

THE BACK OF THIS DOCUMENT CONTAINS A HISTORY OF CHANGES OF NAME AND CORRECTIONS
Before accepting copies, sight unaltered original. The original has a coloured background.
The back of this document contains an artificial watermark. Hold at an angle to view.

REGISTRY OF BIRTHS
DEATHS AND MARRIAGES
MELBOURNE 24 May 2001

I hereby certify that this is a true copy of particulars recorded in a Register in the State of Victoria, in the Commonwealth of Australia.

Registrar

Certificate of the Birth of John Carew Eccles at Jumbunna.

Table of contents:

John Eccles with his parents and sister, Rose, about 1910.

Prologue:

Taking a run to vault high over the bar, the young John Eccles propelled himself upwards on the bamboo pole, over the bar and landed safely. The bar was lifted higher and higher that day and gradually he inched up to become Inter-University pole vault Champion for 1924. It was a height that remained unchallenged for many years. Just as he had with the pole-vaulting, John Eccles continued to set himself goals in his life, rising to the top of the scientific world until, in 1963, he won the Nobel Prize for Physiology or Medicine. In his University life at Melbourne University in the 1920's, he gained many honours and, as a top student, he became the first ex-State High School Student to win a Rhodes Scholarship. Jack Eccles[3] retained his astute brain to the end of his life, always setting goals – books to write, arguments to consider, mountains to climb and countries to visit. He had a restlessness, which kept him going – a need to perform.

In his study of Zoology at Melbourne University in the 1920's, Jack Eccles read *Origin of Species* by Charles Darwin and, although he was attracted to his ideas, he soon realised that Darwin's theory had no explanation for the experiencing self. He decided to devote his life to the mind/brain problem and he stayed faithful to this goal for the next seventy years. In his book, *Evolution of the Brain,* published in 1989, he agrees with the Darwinist hypothesis of biological evolution, but goes beyond the materialistic concepts of Darwinism, "to reinstate the spiritual self as the controller of the brain - begun when hominids experienced self-consciousness".[4] Jack had been brought up a Christian, believing in the soul or self and he kept these beliefs all his life, even when ridiculed by materialist scientists.

John Carew Eccles, Pole Vaulter, 1924

In his life, Jack was not interested just in science but had a diverse range of interests from gardening to mountain climbing, to travelling, and also had a wide appreciation of the arts and culture. He was also interested in environmental issues such as preserving the hinterland of the beaches on the South Coast of New South Wales. He loved to debate his scientific beliefs with those who opposed him and this led to international discussions. Like Einstein, he believed that his responsibilities went beyond his research; in his case, to subjects like freedom of speech, retirement age and belief in immortality.

When he died, there were long obituaries in many newspapers, including *The Times* and *The Guardian*. *The Australian*, Tuesday 6 May 1997, stated "Sir John Eccles stood tall in the front rank of Australian scientists. In a brilliant career of demanding experimental work spanning the decades from 1929 to 1972, he became a giant among the leading international figures in brain research, described as Australia's most distinguished neuroscientist". This was particularly so in 1963, when he won the Nobel Prize for Medicine or Physiology.

---

[3] In the first half of the 20th Century, men or boys with the name John were frequently referred to as Jack. John Eccles was more often than not referred to as Jack, and the names, John and Jack, are used interchangeably throughout the text.
[4] Evolution of the Brain, 1989:xi.

Nobel Prize:

The Nobel Prize, described as the greatest academic honour that can be bestowed, was created by Alfred Nobel (1833 – 1896) a Swedish inventor, chemist and engineer. Nobel made the bulk of his fortune from his invention of dynamite and allied explosives and also his large holdings in the Baku oilfields in Russia. When he died in 1896, he willed that the bulk of the income from his fortune be "annually distributed in the form of prizes to those who, during the preceding year, have conferred the greatest benefit on mankind". The Prizes are awarded in the fields of literature, physics, chemistry, and physiology or medicine. On 10 December each year, the anniversary of Nobel's death, the medals and diplomas are presented by the King of Sweden. The glittering presentation ceremony, held in the Concert Hall, is followed by an elaborate dinner in the City Hall, where the Swedish royal family, academics and government leaders toast the new Nobel laureates The list of scientific winners in the 20th Century include Albert Einstein for Physics, Madame Curie for Chemistry, Lord Rutherford for Chemistry, and Sir Charles Sherrington for Physiology. When Sir John Eccles won the prize in 1963, he was joining a band of the great scientists of the 20th century. He was one of a handful of Australian scientists who had won this great honour up to that time, including Lord Howard Florey and Sir Macfarlane Burnet.

Sir John Eccles shared the prize with two of his colleagues, Alan Hodgkin and Andrew Huxley, for the study of the transmission of nerve impulses in the nervous system.

In his book, *The Neurophysiological Basis of Mind,* he studied the question of how we think physiologically. This has been a question that mankind has wondered about for thousands of years. The ancient Greeks like Aristotle surmised that memories coming from our senses make imprints on our minds like those of signet rings. It has been said of Sir John Eccles that he did not merely surmise how we think but was on the way to discovering the traces of such imprints in the brain through his work on nerve cells. [5]

There are up to 10,000 million of these cells in the brain and millions more in the spinal cord. The cells "pass on information from one to another by 'firing' a chemical message across the synapses, or gaps between them, at the fine, branching fibres called axons which are their extremities. It was Sir John's achievement to record the electrical potential of single nerve cells in the living body by means of incredibly thin glass needles connected to electronic apparatus".[6] It was said that he knew more about the brain and its working than anyone else in the world but, in his usual self effacing way, he said that mankind was only at the beginning of understanding how the brain works.

One day when I (the author) disturbed him in his study, he was looking at some diagrams he had drawn. "Look Mary," he said pointing to diverting branches of nerve pathways. "If you're trying to remember something and it is on the tip of your tongue, you may have gone up the wrong nerve pathway. You need to come back to this juncture here and then go up the other pathway to find the information." It was simply but enthusiastically put and explained his abiding interest in our brains and our minds. His work and the work of his colleagues have helped in many other scientific fields - in psychology, psychiatry, education and of course in the treatment of many neurological diseases like tetanus and Parkinson's and mental illnesses such as schizophrenia.

---

[5] From the Citation at the Presentation of an Honorary Doctorate, Cambridge, 1960
[6] Science and You. Sydney Morning Herald, 20 February 1967:3

Chapter 1:

Early life:

John Carew Eccles was born into a family with both farming and teaching traditions. His paternal grandparents Henry Eccles, from Lancashire in England, and Mary Jane Ingram from Limerick in Ireland had come to Australia in 1849. Meeting on the goldfields near Ballarat, they married in St Francis' Church in Melbourne in November 1850. Over the years, they had nine children and in the 1870's, the family joined a growing number of settlers in the Gippsland area to the east of Melbourne where they established a farm, *Hazelgrove*. John Eccles' father, William, was their seventh child. Although he left school when he was twelve to work on clearing the bush and establishing the farm with his brothers, William was a deep thinker and a poet at heart and he knew that farming would not satisfy him. However, he would probably have remained a farmer had not a teacher, Mr Frank Williamson, come to live with the family. Frank could see that William would benefit from further education and for four years taught him in the evenings. With Frank's encouragement, William eventually applied for admission to the Teachers' College in Melbourne and was accepted in 1887. It was here that he met a young woman, Mary Carew. Mary was the eldest of nine children, many of whom became teachers. Her family, from Tipperary, had also come out to the goldfields where her father had a tannery. By the time William and Mary married in October 1901, William was the head teacher at the Jumbunna School near a mining field in Victoria's Gippsland. As was the rule, Mary had to resign as a teacher the day she got married.

William Eccles, 1866 – 1948.

Their first child, John Carew Eccles, was born on 27 January 1903. He was named John after his forebears on both sides of the family, Carew being his mother's maiden name. His earliest memory was when he had whooping cough and nearly died. His parents took him up to the gas works in Melbourne and held him over the fumes and fire, which was an acceptable treatment in those days. Three times he was held over this to cause him to cough out the accumulated phlegm and he thought if it was done again he would expire from the treatment. His next memory was of the birth of his sister, Rose, at about the same time. William and Mary had only these two children and John, (Jack) was to look back on his childhood as the happiest time of his long life. As a precocious youngster, he could do multiplication sums and could read prior to going to school where his aunt, Maggie, was his first teacher.

Earliest known photograph of John Carew Eccles.

Their head-teacher's house was built in the style of many rural houses in those days - four rooms with a passage down the

Leongatha Picnic about 1908. William Eccles leaning on the buggy, while young Jack Eccles is sitting with his cousins on the buggy.

middle and a verandah on the front, under which was a storage area for fruit from the orchard. Alongside the house was the horse paddock - horse and cart being the main style of transportation. The area was amply watered and a well provided a good supply of water, even in droughts, when they happily shared their water with the poor miners who came begging for a jugful. They lived a simple life in the country with a horse, a cow and some fowls. Living was not expensive, which was just as well as William's salary was only £2/7/0 a week ($4.70 in a direct conversion to dollars). To cut the grass, they used a scythe; they grew vegetables and milked the cow and the only fuel they needed was a bag of oats for the horse when they went out in the horse and buggy. One day his mother, Mary got caught under the clothesline when the horse would not stop.

Young Jack began his schooling when he was 4½ years old at the Jumbunna School, which was a poor building with only three classrooms. His father, who had to teach several classes, initially did not do well on the promotion scale partly because, as a student, he was once rude to a lecturer. When this lecturer subsequently became his school inspector, the poor reports he received kept him in country schools for years. In the end, there were many business and professional men who could testify that he was an excellent teacher. [7] Meanwhile William's brothers, Jack, Dick and Harry had bought their own farm near Leongatha, named *Namoorook*, where Jack and Rose spent many happy holidays.

Koroit:

Early in 1911, William Eccles was transferred to Koroit, near Warrnambool, west of Melbourne. The family travelled by train from Gippsland through Melbourne and onto Koroit, a delightful little town, with the streets set in the usual grid pattern. The buildings in the town were fine examples of Victorian and

---

[7] William Eccles finally got his promotion to the city in December 1919. Promoted from the 4th class to 2nd class, he was transferred to Thornbury State School and the family moved to Melbourne.

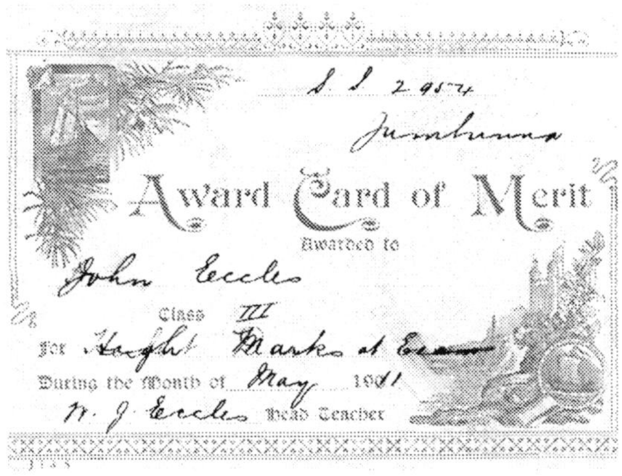

Koroit School No. 618

Eccles Family Home at Koroit, 1911 to 1919.

Edwardian architecture. The area had been first opened up in 1850's for the Irish immigrants who lived in small cottages and grew potatoes, onions and maize in the rich soil.

Although it was only a small country place, Koroit School, No 618, was a minor promotion for William. The school, opened in 1878, was surrounded by a few acres of garden and orchard in the rich volcanic soil. The small headmaster's house, next to the school, was enlarged to accommodate the family. Jack at 8 and Rose at 6 were enrolled at the school and really blossomed under their father's tutelage. The house was adorned with prints of many of the great masters, which William had collected on his world tour of 1901. The bookshelves were full of books won by William and Mary in the Teachers' College: Shakespeare, Milton, Shelley, Keats and illustrated books of great artists, from which William read aloud in the evenings. The family sometimes went to evenings at the Mechanics' Institute or enjoyed meeting with their friends with singing and games. There was also the tennis club where they played tennis, and the bowling club, which was reserved for the men. Although there might be peace and harmony in Koroit most of the time, the world was gearing up for the First World War and emotions ran high. The night that William spoke publicly against the war at the Institute, there was a stone through their window.

Scholarship to Warrnambool High School:

In December 1914, Jack Eccles, aged eleven, won a State Scholarship. Competing against children three years older, he came 19th out of the 20 presented in the whole of Victoria. Over the period William was at Koroit, 5 of his students obtained State Scholarships, two of whom were his children, Jack and Rose.

Jack spent the next four years at Warrnambool High School and, without much effort, kept getting top place in his class. His father warned

The five state scholarship holders from Koroit. From L to R, Iris Taylor, Rose Eccles, Jack Taylor, John Eccles and Enid Stevenson,

Warrnambool High School, 1918, when both Jack and Rose were enrolled.

him "you're top now but just wait until you meet the rest!" Each weekend he would ride his bike home to Koroit or catch the train and during the week boarded with a family near the High School.

While at Warrnambool High School, between 1915 and 1918, Jack learned to be more independent. There was a shortage of teachers because of the war so he taught himself chemistry, did the experiments, took the exam and was $16^{th.}$ in the State of Victoria. A Mr Scarth taught geometry and physics. At first, Jack did not like English as he found it too dry, but a Mr Clark taught him one year and gave him a feeling for language when he recited the *Ancient Mariner*. From now on he began his inexorable rise to the top on the education scale.

Melbourne High School:

In December 1918, William and Mary visited Mr Joe Hocking at Melbourne High School with a view to enrolling Jack for his final year of schooling. They were also hoping he could gain a senior scholarship. It was a time when Melbourne High School secured almost one-quarter of the forty scholarships available to Victoria after the Leaving Certificate. At first Joe Hocking was very dismissive, "We don't just take anybody", he told them. However, when he was told that young Jack had just gained a second class and two third classes in Leaving Honours, he changed his mind magically. "Of course we'll have him. Sign here." So Jack, the young country boy, began his final year of schooling at the Spring St. Branch of Melbourne High School in February 1919.

Sixty or more years later, he remembered:

> As I get older I think more of the past and particularly the very happy and successful year I spent at Melbourne High School in 1919. Poor as were the facilities at the Branch, it was such a happy place. As teachers we had "Nessy" Nesbitt, "Toggy" Graham, and Kate Flynn. Alfred Nesbitt was an Oxford man. He had been headmaster of a Boys' Grammar School at Toowoomba and later became tutor in maths at Trinity College, University of Melbourne. "Nessy", was a brilliant mathematician, a devoted amateur musician and a considerable force in Melbourne's musical circles. He died prematurely in July 1926. This man, Nessy, gave great service to the School, ungrudgingly, even during periods of ill health, and was respected by both his fellow teachers and his pupils who, by his assistance, gained many brilliant successes in University and other examinations. He called me "The Ecclesiastical Gentleman" to the amusement of the class.

The Old Unicornian mentioned that Jack Eccles won the pole vault and the 120 yards sprint in October 1919 at the School's annual sports at the Fitzroy Cricket Ground. It was the first year of peace, and the principal, Joe Hocking, marched his School down to the Masonic Hall where his pupils heard their principal "expound the significance of peace ---- Bare headed and in silence, the assembly stood for a moment to honour the un-returning brave".[8]

World War I was over and the world had become a safer place. In 1919, Jack and Rose and their parents again spent Christmas holidays at *Namoorook*, raking in the hay for uncles Jack and Dick, playing tennis on their court or picnicking at Inverloch Beach just like the old days. One day, Uncle Dick arrived back from the Post Office with the newspaper and Jack thumbed through it anxiously

[8] The Old Unicornian, September, 1981.

John, Rose and their parents en route to Tasmania in 1921.

looking for his Matriculation results. He was staggered to find he had shared the Exhibition in Geometry (State Prize). He got 3 Firsts in Victoria, sixth in Physics, eighth in Algebra and overall First Class Honours. His parents, uncles and sister, Rose, were very excited for him.

It was in this instant that the possibilities of his future lay open to him. He had done well in country schools, but was always hesitant about the competition from the wider community. Now, with these matriculation results, university life lay ahead. If he did well there, who knows, the world was at his feet!

But life must go on: Uncle Dick went out to check the cattle; his parents were getting ready to go bowling; the housekeeper, Mary Hogan, was cooking in the kitchen beside the sleep-out where he had his bed. It was a lazy summer day and the flower gardens were bursting with summer daisies and hydrangeas. The cicadas shrilled in the large gumtrees beside the courts as he served his game against Rose. As he played, Jack could not resist a great feeling of elation and gratitude. He had a loving family who had encouraged him every inch of the way. Life was good. After the tennis they went to watch their parents bowling. That evening, they all gathered around the table for a celebratory dinner.

Melbourne University:

The two fields that appealed to Jack Eccles were Mathematics and Medicine. Mathematics appealed to his imagination, but it might not lead beyond being a secondary school teacher, whereas Medicine could lead to many areas of endeavour. His dilemma was simply solved when he discovered at the Registrar's office, where one enrolled, that the Mathematics course had already started one week earlier, and that he was just in time for Medicine. So he chose Medicine and kept his mathematical love for enjoyment and private study. It was a real turning point because in those days in Australia, a mathematician could never have become a neuroscientist.

It was a choice of career, which led him to becoming a world famous scientist and Nobel Laureate. It was during his first year at University that he studied Darwin's *Origin of Species* and, although the theory of biological evolution seemed logical enough, he came to realise that Darwinian evolution had no explanation for him of the self. So he started on a study of what the great thinkers of the past had written on this, and finally decided, as a medical student, to devote himself to the study of the mind/brain problem. It was after reading Charles Sherrington's book, *The Integrative Action of the Nervous System,* that he decided he would like to go to Oxford and work with the great man.

For the 1921 Christmas holidays, the family had a holiday in Tasmania with friends. Having just finished the first year of his course, Jack wondered how he had fared. Then he received a three-page letter from his lecturer in Chemistry, Dr Rivett, praising his paper. This was unusual when there were 250 students in the class whose papers had to be marked. Rivett was a wonderful teacher of Chemistry. Needless to say, Jack got a First in Chemistry that year. His sister Rose soon joined him at Melbourne University for William believed in educating women – the future teachers of their children.

All through his life, Jack Eccles was fortunate in having had good teachers. From his earliest years he had had his parents and Aunt Maggie, then the teachers at Warrnambool High School, Melbourne High School and then at the University, lecturers like Dr Rivett. The same would continue to happen when he reached Oxford where his mentor was Charles Sherrington. He always revered teachers and the important part they played in society and in educating the young. In turn, he encouraged young fellow scientists and later his own students in their endeavours. One of these students was Marianne Fillenz, who, many years later, wrote that she considered Jack Eccles her mentor. "It was thanks to him that I came to Oxford and his role in my life is similar to that of Sherrington in his, a source of profound inspiration."[9]

In *The Speculum*, the medical student's journal, the following article sums up Jack Eccles' record at the University as a student at the medical school:

> As a scholar, his record is an amazing one. To leave out the numerous scholastic distinctions of his school days, and trace an outline of his career at the University, we find that in his first year he took three exhibitions and a second-class honour. In his second year he was head Apostle and second Prosector, which slip he recovered in third year by taking the exhibitions in both anatomy and physiology. Fourth year was a lean one. He did not get the exhibition in pathology, but he got first-class honours in pathology and everything else, and all the other prizes and exhibitions there were. At finals, space does not permit of a detailed account of his achievements. Suffice it to say that he shared one exhibition and two prizes, obtained first-class honours, and headed the class list. And the above is but a resume.
>
> On the athletic side we find the same sort of thing. At tennis and athletics, he was no mean performer while at school, and at the University he was a prominent member of the athletic team, broke records for pole-vaulting, and gained his Blue. But even this did not exhaust the energies of this remarkable young man. He was a member of the committee of the University Athletic Club, he was the St. Vincent's Hospital representative on the M.S.S. executive, and last, but by no means least, he was a member of the *Speculum* Board.[10]

Rhodes Scholar:

Jack continued with his pole vaulting and was inter-University Champion for three successive years until he sprained his ankle badly. As he had been good both at sports and academically, it seemed natural that he should apply for a Rhodes Scholarship. Cecil Rhodes had provided for scholarships in his will of 1902, for young men who had high mental, moral, and physical qualities, in the hope that their contact with an English University would be of great and mutual benefit. These scholarships were of three years duration at Oxford University. Jack Eccles applied for and became the Rhodes Scholar for 1924. It was the first awarded to a student who had been through the State High School system.[11]

In *The Old Unicornian* we read:

> Victoria had sent a representative year by year since 1902 but never until 1924 had a former high school boy been selected. Consequently, it was with pardonable pride and satisfaction that those associated with the School learnt that Jack Eccles had been chosen Rhodes Scholar. His former teachers and friends did not need added evidence of his great ability, sportsmanship and sterling character. All extended to him heartiest congratulations and full confidence that for him, this was only a beginning. How right they were.[12]

The awarding of the Rhodes Scholarship was made at a stylish luncheon at Government House, Melbourne, where Jack sat on the right side of a charming English lady, the wife of the Governor, and tried to project a cultural knowledge. His success with the Rhodes Scholarship even percolated down to the Inspector's report for his old primary school at Koroit. For the year 1925, under the heading 'Instruction' it was noted – "Last year's Rhodes Scholar was a scholarship winner from this school". A success like this had reverberations in many areas.

The next six months were spent as a Resident Medical Officer at St Vincent's Hospital. Here he met his future wife, Irene Miller, who was a nurse. Then came the news that he had been accepted by

---

[9] Personal communication, 2 December 2002.
[10] The Speculum, June 1925: 22
[11] Up to the out-break of the Second World War, 38 of the Australian Rhodes scholarships were interested in medicine and some remained in England after the expiry of their scholarship. Among them H.W.Florey and J.C.Eccles who became Nobel laureates. Florey won the Nobel Prize in 1945 together with E.B.Chain and Alexander Fleming for their work on penicillin. Florey was then the first Australian–born to win the Nobel Prize for Physiology or Medicine. (Australian Science in the Making, 1988:287).
[12] The Old Unicornian, December, 1924.

Magdalen College in Oxford. It was the college of his choice principally because Charles Sherrington was a Fellow there. Jack regarded it as a "transcendent opportunity" and was grateful to Cecil Rhodes for his great foundation. "My great adventure in life started at the end of August 1925 when the Australian ship *Jervis Bay* left Melbourne and I was farewelled by my mother and father and my dear sister Rosamond, who had been my loving companion for 20 years."

John Carew Eccles, Rhodes Scholar, 1924.

John and his sister, Rose, in his Austin.

Photograph by courtesy of the Library of the Physiology Department, University of Oxford.

Jack Eccles with his friend and mentor Sir Charles Sherrington.

Chapter 2:

Oxford, 1925 to 1937:

Jack Eccles headed off to Oxford with an introductory letter to Sir Charles Sherrington from Professor Osborne of Melbourne University. Invited to the Sherringtons' house, he was to enjoy their hospitality and meet many interesting people over the next twelve years. Before beginning the research, which he wanted to start as soon as possible, he was advised by his Magdalen tutor, Dr McKeith, to study for the Final Honour School in Physiology and Biochemistry for two years and then do his Doctorate of Philosophy in the same field for two more years. Later, he realised that this was the only way to progress through the Collegiate system at Oxford.

In November 1925, he wrote a thank-you note from Magdalen College to Mr Searby and the staff at Melbourne High School:

> I feel I must write before the end of the year, and give you a short account of my experiences. I had quite a pleasant voyage over, and broke the monotony of the sea journey by a trip from Colombo to Kandy, and another from Suez to Cairo. They were short trips but very interesting, both from the point of view of natural scenery and also from the experience for the first time of people differing so much from anything I have been used to.
>
> On my landing at Southampton, I came straight through to Oxford, as I was just in time for the commencement of term. Of course you can imagine how excited I was, as I neared that historic spot. I am in residence in Magdalen College, the most beautiful of all the Oxford Colleges. It is truly a wonderful place with its medieval buildings with their ivy-covered towers, and its wide lawns with shady trees and its parks and gardens.
>
> My own rooms look out into a deer park, where a herd of deer graze in their natural surroundings under the spreading oaks. Altogether, the College grounds comprise about 100 acres. There are about 180 residents in the College. At Oxford, there are over twenty such colleges with numerous other foundations of learning, so you can picture what a wealth of architecture there is. Besides the Colleges there are also non-collegiate buildings such as libraries, churches, museums and science laboratories.
>
> I work at the Physiology laboratory. It is a very well equipped place and offers excellent facilities for advanced work. In the College each undergraduate has two rooms. We have most of our meals in the dining-hall, which is a large room with oaken ceiling and oaken panelled walls. It is over 400 years old. Afternoon tea is served in our rooms and it is at this meal that the undergraduates entertain each other.

Magdalen College, Oxford.

From two to four in the afternoon everybody indulges in sport. At present I am training for rowing. I am in our College Freshman's four. We race next week against the other colleges. The Thames is not very wide, and as it is crowded with rowing craft many narrow escapes occur. Lining the river on one side for nearly half a mile are the various college houseboats. The life at Oxford besides being studious is also sporting and social. It is in these qualities that the life of Oxford and Cambridge differs so much from other universities.

Before I say good-bye I must thank you for the travelling rug, which has been very valuable in the very cold weather we have been having lately. [13]

Magdalen College was amongst the most gracious colleges with its high towers and its fifteenth century cloisters and the chapel where the boys' choir sang Evensong each day in lilting tones that echoed along the panelled walls. Dinner in the old hall was a shared meal with scholars and tutors.

Rowing:

Rowing is a sport that has always been associated with Oxford and Cambridge and students were encouraged to join the crews. Jack Eccles participated successfully in rowing with the Magdalen Second Eight in June 1927, just before the final Honours examination. He rowed in the number six spot as the heaviest member of the crew, who were described by *The Times* as the fastest on the river. For their participation, they earned a bumper supper and each got an oar. Jack's was proudly displayed on his study wall in the various houses he lived in during his life.

Magdalen Second Eight Rowing Team. Jack is sixth from the cox.

Travels in the 1920's in England and Europe:

As an undergraduate at Oxford, Jack had to decide how to spend the long summer vacation when the College was closed down. He travelled with Ivan Turner over much of England visiting the cathedral cities and Wales and right up through the mountains to North Scotland. Later he went on a European tour with Jack Barry, a Rhodes Scholar from Queensland. He was struck by the horror of the First World War, which had finished less than ten years earlier.

On the way to Trois Epis, we passed the Linge where a very fierce battle was fought in the war. The French stormed this important ridge for 3 days and despite the loss of numerous numbers of men failed to take the top. Over 60,000 were killed here. The battlefield has been left exactly as it was at the end of the war and from it one realizes most vividly what a terrible place it must have been. ---- Some distance before the Linge we came to a large French cemetery. I took a

---

[13] The Old Unicornian, May, 1926.

photo of the long lines of white crosses. After we passed the Linge, we came to a large German cemetery with long lines of black crosses. The whole country looked so sad and desolate and contrasts so much with the smiling fertile valleys below.

In the summer of 1927, Jack returned to Australia on long vacation. When he returned to England, he applied for a Junior Research Fellowship at Exeter College which would give him five more years at Oxford at £250 ($500) a year, free rooms and dinners at the High Table. Jack wrote to his parents about the interview, which had gone very well. He had spoken with enthusiasm about his research plans but was still rather anxious about his chances as there were other applicants who already had work published. He was elated that night when he received a note from the Rector, Dr Lewis Farnell, of his appointment to the Exeter Fellowship. After he had moved into a suite of three rooms there in the front quadrangle, Sir Charles Sherrington asked him to join his department and he enthusiastically agreed.

Marriage:

On 3 July 1928, Jack Eccles married Irene Miller (Rene) in Oxford. The priest who officiated at their wedding was Fr Ronald Knox, chaplain at Oxford and later to be a famous Bible translator. Jack's parents and his sister, Rose, were all at the wedding. Their first home was at Headington, Oxford.

Research:

Sir Charles Sherrington had been Professor of Physiology at Oxford since 1891 and was by now a man of tremendous influence in the scientific world. Jack Eccles said that working with Sherrington meant you met all the outstanding scientists of the day because they came to see him. Jack mentioned Alexander Forbes an American, "who has come to the labs for several weeks to do some work. He is quite a distinguished worker on reflexes, but in many points not in agreement with our point of view. Forbes is a very unassuming man and does not at all look like a multi-millionaire – his grandfather was Emerson, the writer".

Sherrington was President of the Royal Society and the British Association. His experimental skill was highly regarded and he in turn set high standards for all his co-workers. To be praised by Sherrington was high praise indeed! Sherrington studied the spinal reflexes and through them came to study man reflecting on his nature which eventually resulted in his famous book *Man on His Nature*.[14] He gave the name "synapses" to the functional connections that are made by close contact between nerve cells. "His magnificent contribution to neurology was concerned largely with showing how the reactions of the nervous system could be explained by the integrated behaviour of individual nerve cells, each of which functioned as a unit and exerted graded excitatory or inhibitory synaptic actions on other nerve cells."[15] Jack Eccles regarded Sherrington as the greatest neuroscientist of the age, and gained a great grounding in university life and in research skills from him, skills which he used for the rest of his life.

Jack wrote about his own belief in "a philosophy that has been integral with my intellectual life since my early years as a medical student. I had been confronted by an irreligious philosophy of monist-materialism that I could not accept. ---- In fact from my reading of philosophy I discovered a pervasive ignorance of the brain at the subtle level at which it could relate to conscious experiences, for example, in neurology. Soon I recognised that my task was to be a life-long study of the brain, or neuroscience as it came to be called!"[16] During his time in Oxford he published eight papers conjointly with Sherrington and also collaborated with Ragnar Granit and J.Z. Young on two projects. One of these papers, published in the Journal of Physiology, was *Impulses in the giant nerve fibres of earthworms*.[17] Curtis and Andersen[18] noted the importance of this work: "In 1932, Eccles, J.Z. Young and Granit showed, for the first time, that action potentials were conducted in both directions along

---

[14] Man on his Nature, 1940, Cambridge University Press.
[15] The Physiology of Nerve Cells, 1957:1.
[16] How the Self Controls Its Brain, 1994:14.
[17] Journal of Physiology, 77,23P.
[18] Both Curtis and Andersen were colleagues of Sir John Eccles in Canberra.

Jack Eccles with sister Rose and their parents outside Exeter College, Oxford, 1928.

earthworm giant nerve fibres, an observation which, when later extended by others to the giant axon of the squid, was of fundamental significance to the understanding of impulse conduction along axons".[19]

Jack was beginning to write scientific papers and wanted to become a good writer with a clear and lucid style. He turned to his friend, J.R.R.Tolkien, Professor of English at Oxford University. In their discussions, they found that they both used Fowler's *The King's English* as the basis for their writing.[20] By the end of his life, about 500 papers had been written by Jack himself, or in collaboration with his colleagues.

In December 1930, he wrote to his parents:

> I went to the Physiological Society meeting in London last Saturday. I went early in the morning and met McSwiney and we spent the morning together, going to the meeting after lunch. I dined with Denny-Brown that night and we had a long talk about the book afterwards. [This book is referred to on the next page]. I arrived in Oxford at 11.40 p.m. and had to walk home (3 1/2 miles because the buses had stopped for the night so did not get home till 12.30.) [Rene added a note: I was waiting on the doorstep with a hot water bottle and his warmed pyjamas].
>
> We physiologists of Great Britain were united by our membership in the Physiological Society, which, I think, was then in one of its great periods. It was distinguished by the critical discussions that followed each paper. These criticisms were often severe but it was an unwritten rule of the society that all criticism had to be accepted in a sporting manner. On no account must it be taken personally.

And again two weeks later:

> Well the papers have gone off at last so now I have more time to myself. I sent them off this morning after a strenuous week's work putting in the final touches --- The whole thing will amount to about 100 pages of print and there is not much of that that is superfluous I can tell you. For the most part it has been written 3 or 4 times. When it was all over Sir Charles told me that I wrote much better English than any colleague he had ever had, and said several other nice things so I am suffering from a swelled head. He had practically no corrections to make to my writing at all. But to write that way is for me a work of infinite patience. ------ Rene's typing has been a great help for I don't think that I could ever have finished the task of copying it all out each time. She has typed the whole thing through at least twice from my pencil notes, which were too rough for any of the typewriting firms to tackle. The papers will be published about March, but before then Sir Charles and I are going up to read them before the Royal Society. On the whole I am very pleased with the whole contribution, and I think that it puts us years ahead of anyone else in the world. --- Of course this is only the beginning of the story and there are immense possibilities for further developments.
>
> I think that it is wise for Sir Charles to take things more easily now. He is almost 74 and it is surprising that he is still so active. I don't know how long he will continue as Professor. I don't think he has any thought of retiring - he is the sort of man that will die in harness. If he carried on for two or three years, it has been suggested that I would be his most likely successor - but don't tell anybody about that. If for less, McSwiney is probably the most likely man. The laboratory at present is in its most flourishing condition and will become more famous than ever when the present papers and the book are published. Well I think that I have talked about work

[19] Historical Records of Australian Science, Vol 13, No. 4:441.
[20] Tolkien became godfather to Jack and Rene's first child, Rosamond.

enough, I am afraid that my mind is rather full of it at present - there is nothing else to do for the weather is foul.

Baby Rose is in her usual good form. She had a Christmas tree and was very excited about it all. She had lots of presents, which she tried to give away to everybody else. Tolkiens came and also the Hoffs. For the most part our Christmas festivities were very homely. We lunched at Sherrington's on New Year's Eve and went to Midnight Mass.

Earlier in 1930, Charles Sherrington had announced that Clarendon Press had persuaded him to write a book on the research of the Oxford School. He accepted the invitation on the condition that it would be a conjoint effort - the chosen team of his associates being Creed, Denny-Brown, Liddell and Eccles. It would be an account of the Sherrington school of thought of the previous decade, particularly on the excitation and inhibition of the spinal cord. Eccles, excited by the prospect of contributing to a book at so young an age, wrote the first three chapters. Eventually, the book, published in 1932, was called, *Reflex Activity of the Spinal Cord*. In 1930, Jack Eccles received the Chapman Prize, worth £20 ($40), for his outstanding research work. Sir Charles had to judge between Denny-Brown and Jack and decided they were each worth a prize. Each got the full value of the prize and it was four years before it was offered again.

Jack and Rene with baby Rosamond, born 1929

His salary as a research fellow and tutor was rather meagre when one considers he was married with two children: £250 ($500) for being a Research Fellow; £150 ($300) Christopher Welch Scholarship; £200 ($400) as a demonstrator; £100 ($200) for tutoring – £700 ($1400) in all. In 1932, Jack was awarded the Rolleston Prize which was a prestigious award given to a researcher from Cambridge and Oxford for outstanding research in the whole field of biological and medical science. The prize money of £125 ($250) was used on renovation of the house and installing central heating in part of it.

Apparently he was better off staying in Oxford than returning to Australia at this stage. He wrote to his parents that it would take years to duplicate the apparatus that he had in the Oxford labs. Furthermore it would cut him off from other researchers, particularly Sherrington. In addition, he was just beginning to make an impression on the Royal Society. "I am in the midst of preparing for the communication to the Royal Society tomorrow. I will have to talk to them for about 3/4 hour on the work that Sir Charles and I have been doing and then put up with questions that may be asked." Where else could an aspiring young scientist find this incentive and encouragement in his life?

According to Masao Ito, Jack Eccles learned a great deal from Sherrington, "about the classic concepts of reflexes, such as the convergence and divergence of synaptic connections, excitatory and inhibitory states in a pool of motor neurons, and the integrative action of the central nervous system"[21]. When the news came through in October 1932 that Sherrington had shared the Nobel Prize for Physiology or Medicine with his good friend Dr E. D. Adrian of Cambridge for "their discoveries regarding the function of the Neurone", everyone was wildly excited. The scientists and wives arrived

[21] Nature Vol 387, 12 June 1997.

at the Sherringtons' house for a surprise party. It was a glorious climax for his life as a brain scientist and for his associates. They drank a toast to Sir Charles and Lady Sherrington in sherry.

In the early 1930's, Eccles realised that the future of neuroscience belonged to electrical recording. Dr Matthews in Cambridge had designed a mechanical oscilloscope with mirror recording, which he and Dr E. D. Adrian were using successfully. In 1931, Sherrington bought an oscilloscope for Jack's lab to replace the string galvanometer he had been using. With the new Matthews oscilloscope, Jack was able to record electrical responses of nerve fibres and neurons. However the machine was temperamental in its recordings and he wanted to contact Dr Adrian for help. At first he was a little hesitant to approach the Cambridge scientist who had been his examiner for his Doctorate of Philosophy. Not only did Adrian invite him over to Cambridge, but also met him at the station. That evening, they attended dinner at Trinity College where he was seated at the high table between two famous scientists, Rutherford and Eddington! The problem of the new machine was soon solved when they realised that the machine was too cramped in its assembly to provide good results.

George Cox organised the animals for the labs and supervised the technicians who were recruited from the local secondary schools. One young technician, Meadows, was comforted by Jack when he fell under the ire of George Cox. Later Meadows, while still a technician, became the Lord Mayor of Oxford and entertained the Physiological Society of Great Britain at a reception in the Oxford City Hall. Jack often said the great success of Meadows reminded him of Dick Whittington! Another of the young technicians was Arthur Chapman who accompanied Eccles on all his odyssean journeys around the world until he left the Australian National University.

To provide complete electrical silence, Jack Eccles built a shielded room and made an elaborate design for the door to complete the shielding. Peter Eccles remembered seeing his father's laboratory:

> --- he had a very interesting method of doing things. He had a Faraday Cage in his laboratory. Made out of copper mesh stretched over a big wood frame to keep unwanted signals and voltages out. To get inside, they walked through a doorway, which he carefully shut. Then he had a weight on a stiff pendulum arrangement about 60 cm diameter that swung a set of contacts through other contacts. At a specific time after he released the weight into its swing, it released a linear paper chart to travel at a constant speed. This way, the contacts on the pendulum arrangement stimulated a live nerve fibre, and set the chart to record whatever his experiment required from another set of contacts on the nerve fibre. The weight was caught at the top of its swing and could be set to run again as required. The whole thing looked like a Heath Robinson set-up. ---- As a boy I was most impressed with this great arrangement, but I also met his staff who looked after this elaborate mechanical timer, and prepared his nerve fibres. Arthur Chapman was his faithful technician through most of Dad's experimental career.[22]

Peter, Alice and Rose.

Jack Eccles wrote an account of his first experimental success when he demonstrated that previously held results of research on the angle of the isometric muscle twitch were due to the friction in the bearing of the recording myograph. The angle was an established dogma at Oxford, but when Jack constructed a new myograph, it failed to deliver the angles! It had been the fault in the old myograph that had given the wrong results previously. His discovery and new findings led to his addressing the Physiological Society in December 1929 when he was 26 years old.[23]

By 1935, the Eccles family had grown to three children and, each year, the family rented a waterfront house in Coverack Village in Cornwall. In the nearby cove, the children, Rose, Peter and Alice, frolicked in the sand while Jack escaped from the pressures of research with a collection of books.

[22] Description by Dr Peter Eccles, personal communication, 1999.
[23] My Scientific Odyssey, 1977:2

Sherrington loved the children who often picked posies for him.[24] And he was always supportive of the families of his researchers. He sensed that Jack Eccles worried that Rene's qualifications as a nursing sister did not put her in the same category as many of the other wives at Oxford, but Rene had an ally in Charles Sherrington as shown in this letter, written 26 December 1935, after he had shifted house.

> Dear Rene,
>
> How good of you and Jack to send me these splendid gloves and they fit me to perfection: you are both very clever. Thank you very much indeed. How are the children and how did they get through Christmas: prima, secunda, tertiaque! I have now been nearly 5 weeks in this new little house. Tell the children I have lots of wild rabbits in the garden. I have enjoyed Jack's paper in the new number of the Journal. My love to the children and if I may say so to you all, Yours C.S.Sherrington.

Thinking back on his Oxford days, Jack Eccles wrote:

> England, in my thirteen years there (1925-1937), was a delightful and stimulating place for a young academic, although by present standards the laboratory facilities were primitive. There were almost no research grants and no secretarial assistance even for Sherrington. We had to type our papers and service and organise our equipment, which gradually became more complicated with string galvanometers giving place to cathode-ray oscilloscopes in 1933 and valve amplifiers. But in research, the competition was not severe. The world literature was unbelievably small, so that one could easily survey the total publications, not only on the nervous system, central and peripheral, but also on all types of muscle and all types of sensory systems." [25]

Jack Eccles' research in Oxford led him to believe that transmission at synapses between nerve cells was an electrical process. In contrast, the rival theory, advanced by Sir Henry Dale, was that transmission was primarily chemical. This controversy had the effect of defining problems and stimulating research.[26]

A succinct description of these problems is given in *Australian Science in the Making*:

> The mechanism whereby a nerve impulse is transmitted from a nerve ending to a muscle across the neuromuscular junction or from a nerve terminal to another neurone across a synapse in the central nervous system was the subject of considerable controversy during the first half of the present century. Of two theories, the work of Sir Henry Dale and his colleagues in London supported the hypothesis that a chemical substance was responsible for the passage of the impulse from nerve ending to effector cell. The second or electrical hypothesis, however, had to be tested; Eccles set out early in his career, while still at Oxford, to do this. His arguments and discussions frequently brought him into friendly, but sometimes heated, conflict with Dale and his supporters.[27]

There were many arguments about these differing views at the Physiological Society meetings, the Minutes of which were for general distribution. These friendly scientific arguments and the critical discussions may have had far reaching consequences that the participants never dreamed of.

During the 1930's, with the impending war looming, German military scientists were apparently experimenting with nerve gases, which could cause the death of thousands of people if released. They followed the discussions in England on the nervous system, particularly among Dale, Eccles and others and decided that the English scientists knew more about the effects of nerve gases than they were admitting.

---

[24] Sherrington's wife died from cancer in 1933.
[25] My Scientific Odyssey, 1977:3.
[26] Eccles was viewed by the science establishment in the United Kingdom as a "young Antipodean whippersnapper contradicting the great man [Sir Henry Dale]". (Professor Max Bennett, quoted in the Medical Observer, 11 July 1997)
[27] Australian Science in the Making, 1988:303.

Anthony Tucker, writing in the Guardian in May 1997, said:

> Although it is often suggested otherwise, the bitter debate [between Dale and Eccles] did not obstruct scientific progress. Rather it defined and stimulated specific areas of research which laid the foundations for a later and much deeper understanding of the biochemistry of nerve signal mechanisms. The connection with nerve gases is strong, but whether the arguments had an important bearing on the course of the Second World War remains uncertain. Dale's theory of chemical transmission and the counter arguments of Eccles had been widely published and discussed by 1936, the year in which Gerhard Schrader of I G Farbenindustrie identified, among new potential pesticides, one of enormous lethality. This appeared to work by producing rapid death through muscle paralysis. Its potential as a weapon was drawn to the attention of the German military and in 1937 it emerged as the first nerve gas, Tabun. Very rapid in action and far more lethal than anything before, it possessed probable war-winning capabilities against an unprepared enemy.
>
> There was little fundamental research in Germany on neural signal transmission but German military scientists, monitoring scientific literature for research progress, were very much aware of the vigorous controversy in Britain about the chemical transmission of nerve impulses. German scientists were fairly sure that British scientists, of the calibre of Eccles and those in Dale's group, must know something about nerve signal blocking agents such as Tabun, but there was a total absence of papers on this new class of compounds in open scientific literature, both in Britain and the U. S. This gap in the literature was misinterpreted by German scientific intelligence as censorship, indicating, wrongly, that Britain was already aware of the nerve gases.
>
> So Germany's potentially devastating stockpiles of several thousand tons of nerve gas were never unleashed because Hitler mistakenly feared that Germany would suffer reprisals in kind from the allies. The allies stumbled on the gas plants and stockpiles only later studying them and realising their significance. [28]

If this was so, then scientists like Dale and Eccles had a great effect on the outcome of the war, unwittingly saving thousands of lives.

In 1936, the question arose of a successor to Sir Charles Sherrington in the Chair of Physiology at Oxford. It was considered that Eccles was too young at 32 and an older man was elected to the position. Jack, unhappy at missing the Chair, had to look elsewhere for employment. Previously, he had been offered professorships in three other universities between 1932 and 1935, but it would have meant leaving Oxford at a time when his tenure there was still solid and he had better equipment in his labs than found elsewhere. One of these was Sheffield University where Howard Florey, who was Professor of Pathology, was keen for Eccles to take the Chair in Physiology.

But with the Chair in Oxford being lost to him, Eccles, in 1937, accepted the Directorship of the Kanematsu Institute at Sydney Hospital and embarked on the next part of his Odyssey, never again to accept a position in his beloved England. He had only to "exercise general supervision of the clinical sections of the Institute that were fully housed in the three lower levels of the fine new Kanematsu Building". He was gifted with great energy and enthusiasm as well as a wide range of research experience for his challenging new position. [29]

Sherrington wrote on 31 January 1937

> Dear Jack,
>
> I am indeed sorry at your decision. I only trust you may be wise in taking it – and that the new position will be all you wish. No doubt you would be more completely a master of your time. But I can't help thinking of the loss to Oxford and to all your many friends here. I have always heard that Sydney is a fine place, wonderfully situated and a perfect sun bath for the children. My love

---

[28] The Guardian, 7 May 1997.
[29] My Scientific Odyssey, 1977:3.

to you all. Rene must feel excited at this new horizon. Granit for one will be sad to lose you but not more so than I am. Yours ever, C.S.Sherrington.

Sherrington wrote again on 2 April 1937:

> ---- we shall miss you and many will do so besides us. However you will form your own group and the Southern Cross will smile on you and we shall have the greater exchange of young people and new ideas travelling back and forth. It will be a great thing too for the children. I hope they do not forget the Oxford they began in. It's a bad thing for Magdalen – what will they do? Yours ever, C.S. Sherrington.

As can be seen, the reasons for his leaving England were many and varied. If Jack had obtained the Oxford Chair, he would have been tempted to stay on and weather the uncertainty of the political scene. However, with Sherrington retired, the focus of the research in the department changed completely and he lost interest in continuing there. The threat of the looming war with Hitler forced him to look further afield for employment. So Jack and Rene returned to Australia with their young family of four children – Rose, Peter, Alice and William. Jack 's parents and sister in Melbourne were elated at the news but Rene and Jack were sad to leave their friends behind in Oxford. The trip was quite pleasant on board the *Orford*. He wrote, "We passed south of Spain and could see in the mountains signs of the Spanish Civil War that was still on. As we passed through the Suez Canal, we felt relieved to have left behind Europe with the ever-growing Hitler threat". Also on board were Joe Lyons, Prime Minister of Australia, and his wife Enid, who had just been to England for the coronation of George VI.

Sir Charles Sherrington.

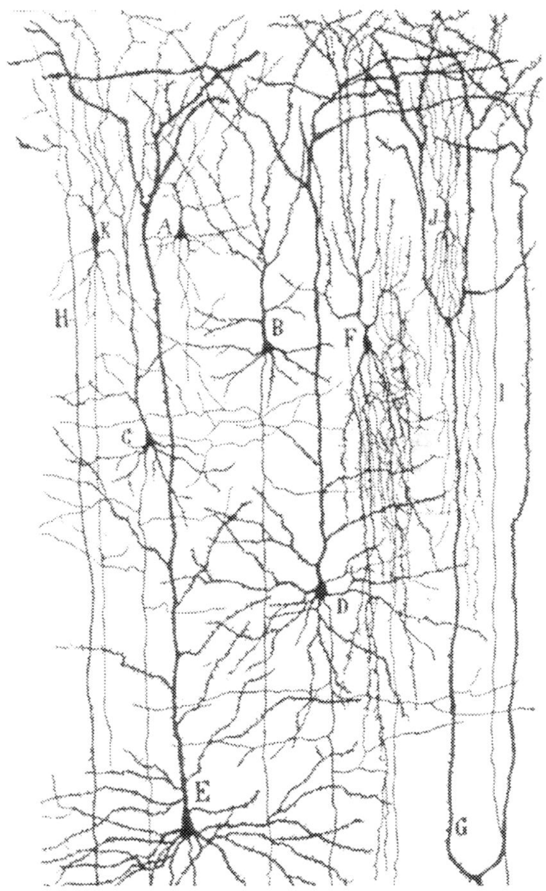

Neurones and their synaptic connections. Eight neurones are drawn from a Golgi preparation of the three superficial layers of frontal cortex from a month old child. Small (A,B,C) and medium (D,E) pyramidal cells are shown with their profuse dendrites covered with spines. Also shown here are three other cells (F,J,K) which are in the general category of Golgi Type II with their localized axonal distribution. G – I are dendrites from deeper lying neurones, G being the apical dendrite of a large pyramidal cell. (Ramón y Cajal, 1911). (As shown in J.C. Eccles "Facing Reality", 1973:10)

Steve Kuffler, Jack Eccles and Bernard Katz, scientists of the Kanematsu Institute, Sydney.

Chapter 3:

Sydney, 1937 to 1943:

Because he was anxious to see his new working environment, Jack Eccles visited the Kanematsu Institute soon after his arrival in Sydney and met Dr Ritchie, the Senior Physician of Sydney Hospital, who became his friend and adviser throughout his Sydney days. He found that the Director's office had a wonderful view of the harbour but could see some changes were needed to the structure of the laboratories. An architect was appointed to prepare plans for these necessary changes. Arthur Chapman, his experienced technician from the Oxford Laboratory, had also accompanied him to Sydney.

In Sydney, in the 1930's, there were two private institutions that fostered research: firstly, the Kanematsu Memorial Institute of Pathology, attached to Sydney Hospital, established with financial assistance from the Japanese firm Kanematsu (Australia) and secondly the Kolling Institute, attached to the Royal North Shore Hospital, named after Mrs Eva Kolling, the first benefactor. Dr Rudy Lemberg, a biochemist and refugee from Hitler's Europe, began doing research at the Kolling Institute in 1935 and continued until he retired in 1972. He became a lifelong friend of John Eccles who tried to build up a group of friends from among the fellow researchers in Sydney.[30]

Jack and Rene bought "Dunsfold", at 14 Clanalpine St Mosman for £3300 ($6600). It had beautiful views over the harbour and a lawn tennis court. They found an architect to renovate the house, add a double garage and do some internal remodelling. The downstairs front rooms were really large and had a wonderful view of Sydney Harbour to the south. The room to the right was Jack's study and contained his prize display, the engraved oar commemorating his "Oxford 8" days, and the HMV 78 rpm record player, mostly for 12 inch orchestral pieces.

The garden had a shed that became known as the "Hobbit House" in honor of the main character in Tolkien's new book. Swings, parallel bars and a sandpit were built in one corner of the garden. Then there was the vegetable garden, plus rear lawn. But Jack Eccles was aware of the downside of the move from Oxford:

> Sydney was a lovely place to live, but the academic isolation was severe. The Institute I was to direct was simply the routine pathology department of a large general hospital in the city some three miles from the University. Nevertheless, with good help from the Institute Committee, and the Hospital Board, I was able to construct research laboratories on the top floor utilizing, for a start, the equipment that Professor John Mellanby had kindly allowed me to bring from my two research laboratories at Oxford. I decided to study the electro-physiology of neuromuscular transmission in muscles of the cat hind limb because I thought that it could lead to results of clinical interest.
>
> The academic wilderness soon blossomed. In 1938, Stephen Kuffler arrived as a refugee from Austria, and by good fortune I heard of this young pathologist in search of a position. So he became a neurobiologist in the Kanematsu Institute with almost no background knowledge of the nervous system! I managed to attract Bernard Katz from England on a Carnegie Fellowship. Thus in this city, through the machinations of Hitler, we three were sheltering securely in remote Australia and studying neuromuscular transmission in cats and frogs.[31]

Jack Eccles had first heard about Stephen Kuffler from his friend Fr Richard Murphy. The young Austrian doctor had been a leading anti-Nazi in the Medical School in Vienna, and, managing to avoid arrest, he escaped to Hungary the night the Nazis entered Vienna. Now in Sydney, he had been trying to get a University position in Pathology. Eccles was delighted to hear that Steve was a researcher and also had twice been the Austrian Junior Tennis Champion. Steve and Fr Richard Murphy were invited to a tennis afternoon and this proved an ideal meeting. Subsequently, Stephen was appointed

---

[30] Australian Science in the Making, 1988:297-298.
[31] My Scientific Odyssey, 1977:4.

an associate at £500 ($1000) a year to the Kanematsu Institute. He found a unit in Mosman and settled easily into life in Sydney.

The next to join the team at the Kanematsu Institute was Bernard Katz whom Eccles had met at Physiological Society meetings in 1936 and 1937 in England. He was recommended by Dr Feldberg who said he was the best of the refugees from Europe for the position of research assistant. He arrived on 14 October 1939 with his parents and became the third researcher at the Institute.

> While at the Kanematsu Institute, Eccles and his colleagues, Katz and Kuffler, tested the electrical hypothesis concerning transmission across the neuromuscular junction. Eventually, in 1949, they reached the conclusion that the impulse was transmitted by the chemical transmitter, acetylcholine.[32]

Each morning, Jack Eccles, Bernard Katz and Steve Kuffler travelled by ferry across the harbour to work on their research projects or lecture to Medical students at Sydney University. From 1938 to 1940, Eccles gave 6 lectures to the third year Medical students at Sydney University on general physiology of nerves and muscles, and lectures to the fourth year students on neuro-transmission. One of the papers the three scientists produced conjointly was *Nature of the 'end-plate potential' in curarized muscle.* [33]

If they worked late, the three of them would have dinner at an Italian restaurant, then dash down Sydney streets to catch the last ferry at midnight. Lunches would be eaten sitting on the grass at the nearby Botanic Gardens, and at weekends they played tennis on the grass court at 14 Clanalpine Street. A favourite beach was nearby Sirius Cove, where Peter Eccles can remember Steve Kuffler swimming strongly towards the middle of the harbour totally oblivious to the shouts of "Watch out for the grey nurse sharks".

In those years, Sydney had an excellent orchestra that performed often in the Town Hall; also there were frequent recitals and chamber music concerts in the Conservatorium of Music in the Domain quite close to the Kanematsu Institute. In the Independent Theatre, North Sydney, there was a very good Repertory Theatre group producing interesting plays. Among their friends was an Austrian sculptor Arthur Fleischmann who was introduced to them by Steve Kuffler. Later, young John Eccles Junior, at about two years old, posed as a model for Fleischmann who was sculpting a statue to "Child Beauty". This was to be a memorial for the son of then Governor General of Australia, Lord Gowrie, for the Garden of Government House, Canberra. This statue still stands there in the rose garden.

Eccles Home, Sydney, 1937 to 1943

Rene and Jack were closely associated with a Dutch Catholic group, the Grail, who believed in making religion a source of joy to the young. The Eccles family often visited their weekend religious festivals, and, in return, taught them English Folk Dances learned at Oxford. Steve Kuffler often came along and there he met Phyllis Shewcroft, a medical student. Jack was the Best Man when they married in St. Mary's Cathedral.

The war was coming closer to Australia and to Sydney in particular - a Japanese attack was expected at any time. There were blackouts at night and inspections. Cars had to have hoods over the lights with slats

[32] Australian Science in the Making, 1988:303.
[33] Journal of Neurophysiology, 4:362-387.

to focus the light on the road only. Parades of the soldiers of the 6th and 7th Divisions were held down the main streets of Sydney. They had returned from the Middle East and were now on their way to New Guinea and Darwin to defend the country against the advancing enemy. Eccles, a scientist, was not called up, but worked on wartime experiments such as artificial fogging. From 1941 onwards, as the war increased, the Kanematsu Institute became the Australian centre for blood serum preparation and for applied research on such acoustic problems as noise protection and communication in the high noise levels of tanks and planes. Bernard Katz was chosen to become a radar expert.

In the house at Mosman, the Eccles family had a bomb shelter, which was hardly used during the war. One occasion it was used was on 31 May 1942, when two midget submarines managed to penetrate Sydney Harbour. They did little damage, sinking an out-of-commission ferryboat, but many servicemen did lose their lives.

Jack and Rene Eccles in 1940, with Alice, Mary and, William, and, in front, Rose and Peter.

Fellow of the Royal Society 1941:

In 1941, Jack was elected to the Fellowship of the Royal Society. Sherrington had submitted his name as early as 1935 but it was not until March 1941 that he received a telegram of congratulations, "Heartiest congratulations Royal Society Election, Adrian". To become a Fellow of the Royal Society was a very great honour indeed. To be able to write the letters F.R.S. after one's name meant that one's research was recognised as being of first-class standard. Jack was now 38 years old and the work he had done with Sherrington had led to this after an eight-year wait. There was a great celebratory party at the Mosman house. Jack was pleased to hear that his former Chemistry lecturer, Dr (later Sir) David Rivett, was elected that same year.

Eccles, Katz and Kuffler leave the Kanematsu:

In the early 1940s, Katz, Eccles and Kuffler did wonderful research in the Kanematsu Institute and it was thought they would continue after the war. However, events had begun to be set in motion that would change the status quo. First, changes were made to the laboratories, which Eccles had wanted to extend - the Hospital Board needed the space for residents, and other things made it impossible to continue. As a result, in 1943, Eccles accepted a post as Professor of Physiology at the University of Otago at Dunedin, New Zealand. His technician, Arthur Chapman, helped pack up the apparatus at the Kanematsu Institute, and went with him to his new position.

Bernard Katz and Stephen Kuffler, both later intellectual giants in the area of neuroscience, left for overseas postings. Katz later went to AV Hill in London and Kuffler went to Chicago. Katz later won the Nobel Prize for Physiology or Medicine in 1970 and was elected to the Royal Society in 1952. Kuffler, who was elected a Foreign Member of the Royal Society in 1971, had a distinguished career at the Harvard Medical School.

The Medical Observer commented:

> The three researchers were destined to be acknowledged universally as leaders for their pioneering research on how nerves at the synapse control muscle cells. Professor Bennett believes it was a tragic loss for Sydney University when the formidable trio left Sydney to take

up new appointments elsewhere. They had been a familiar sight on campus, where all three lectured to medical students in the Listerian Theatre of the Anderson Stuart Building.[34]

Jack and Rene now had seven children as Mary, John and Judith were born in Sydney. It was sad for them to leave their friends and sell the lovely house in Mosman. Because of wartime restrictions, the house had to be sold for what it had cost plus the cost of any renovations.

Left: Otago University, with its distinctive Clock Tower.

Below: The Lindo Ferguson Building, which houses the Medical School, where John Eccles worked.

Photograph of the "Clock Tower Building" by courtesy of the Otago University.

Photograph of the Lindo Ferguson Building by courtesy of the photographer, Nairn Smith.

[34] Medical Observer, 11 July 1997.

Chapter 4:

Dunedin, 1944 to 1951:

> And so my odyssey continued to the remote university in the south of the South Island of New Zealand, the closest university to the South Pole. Even there, I found stimulating challenges and the opportunity to develop scientifically.[35]

The family travelled from Sydney to Melbourne to meet their parents and grandparents, William and Mary Eccles, at Kew and their cousins. Within a week, on 1 January, 1944, and in great secrecy, they boarded the military transport ship, *Moaltan*. There were few passengers and most were troops returning home after action in the Middle East. As Japanese submarines had been sighted, gun practice was a regular occurrence.

They arrived in Dunedin and, within a month, Rene and Jack bought the house at 84 Cannington Road, Maori Hill. It had an outlook onto the hills, Mount Carghill in front and Mounts Flagstaff and Swampy on the sides. Often snow-covered in winter, Flagstaff and Swampy were both climbed by the family. It was a matter of honour to climb the peaks that you saw so often. Friday nights were folk-dance nights when university friends and neighbours joined in for a good time. There was also the three-quarter-size tennis court at the side of the house for tennis afternoons.

Following their parents' training, both Jack and Rene were keen gardeners with a large area under cultivation with all manner of fruit and vegetables. Canes of black and red currants, raspberries and gooseberries grew down the back of the vegetable garden and were a favourite place for the hide-and-seek games which the children played in spring and summer amongst the flowering rhododendrons. Evening meals were eaten in the large kitchen, with all the family around the large table. Jack usually dominated the conversation and talked science or physiology with those who could understand, or he would spend the mealtime giving the family a geography lesson, getting a geographic magazine or an encyclopaedia out and explaining different things. He had a great wonder about the world, which he had gained from his father and tried to hand on. After dinner, he would retire to his study where he would work while listening to classical music.

In Dunedin, the family enjoyed many activities, which included trips to the Botanic Gardens where the family went for walks in azalea time; trips to the beach at St Kilda; bushwalking and mountain climbing and picking rosehips out in the fields. There was plenty of work to do and the children were all allocated jobs. In Dunedin, the last of the nine children was born. From oldest to youngest they were Rosamond, Peter, Alice, William, Mary, John, Judith, Frances and Richard.

At the end of each year, Jack and Rene would go on a camping trip, travelling by car to some of New Zealand's most beautiful lakes and mountains, including Mt. Cook. They never lost that spirit of adventure and love of camping and exploring. The family were left in the charge of Rose and the older siblings, which was sometimes rather difficult for them.

Someone once described the sight of the family appearing at Church:

> I had to tell you about the lovely sight of Dr Eccles and his entire family at Mass last Sunday. Mrs Eccles came first carrying the baby (Richard about two weeks) then a small girl and a boy and then the eldest, Rosamond and four other girls and boys. The Doctor carried the second youngest, nine in all. They took up the whole seat. It was really a picture. I would have loved to have taken a photo.

---

[35] My Scientific Odyssey, 1977.

Professor of Physiology, Otago University:

The Otago University was founded in 1869 by the Presbyterian Church with four professorships, Philosophy, English, History, Physics. The Medical School was established in 1876 when Dunedin was the largest city in N.Z., and also the commercial centre because of its gold wealth. At that time it was the only medical school in the country.

> After my ten years of teaching at Oxford, I was enthusiastic to attempt a similar program of years of lectures, practical classes, and discussion classes despite the greatly restricted facilities. In this I was fully supported by Norm Edson, an inspired teacher of biochemistry in the modern form that he had learned from Krebs. So the medical students were subjected to this intensive modernization of physiology and biochemistry. In my first year in Dunedin I lectured to the second-year medical class in the whole of physiology, 75 lectures in all, and I also did much of the first-year course. --- I had to spend many hours each week learning the whole of modern physiology, so that I could lecture on it with authority.[36]

Jack Eccles with his colleague, Ralph Gerard and children Alice, Mary, Richard, Frances, Judith and John.

At the end of his last lecture an unheard of thing happened: the medical students' representative got up and thanked him for all of the effort that he had put into the course and also the staff for their herculean efforts. Jack was pleased at the appreciation of his course, particularly as his research suffered in the face of his teaching load. But then, in the world at large, because of the war, research had virtually come to an end except for a few who fortunately had been sheltered.

Jack Eccles had brought much of his research equipment with him from Sydney and set up the new research room with the help of Arthur Chapman and a local technician, Charlie Morris. Later Jack Coombs helped with his great electronic skill. Because of the teaching load, not much research was done in 1944 but by 1946, Eccles was concentrating on synaptic potentials of the cat spinal cord, following his Sydney work with Katz and Kuffler on the stellate ganglion. In 1946 he produced a paper, *An Electrical Hypothesis Of Synaptic And Neuromuscular Transmission.* Eccles tells the story of how, until 1945, he had been following the conventional scientific method of careful and methodical collection of data from which an hypothesis was formed. This was the inductive method – however, if the scientist's hypothesis is proved wrong, then this is a disaster for him and he could lose his credibility as a researcher.

In doing his research in New Zealand, Eccles said, "I had long espoused an hypothesis which I came to realize was likely to have been scrapped, and I was extremely depressed about it". His research was proving his theory about electrical transmission was wrong. It was at this point, in 1944, that he met Karl Popper in nearby Christchurch and learnt from him that it was not scientifically disgraceful to have one's hypothesis falsified. In fact, "I learnt to formulate my electrical hypotheses of excitatory and inhibitory synaptic transmission so precisely and rigorously that they invited falsification".[37] Eccles was delighted to find an academic of Popper's calibre in such close proximity and invited him to give five lectures at the University on the philosophy of science.

> They were an enormous success among the staff and student body, and there were also two special seminars, one to physical scientists, the other to biological scientists. Many people, including myself, had our scientific lives changed by the inspiring new vision of science that Popper gave us. Briefly the message we got in those memorable lectures was that science is not inductive, but deductive. A scientific project starts as a problem, for example with a theory

---

[36] My Scientific Odyssey, 1977:5.
[37] Facing Reality, 1970:105.

that appears deficient or inadequate. New hypotheses are developed and tested experimentally then either to be falsified, or corroborated, but the claim of verification should never be made.[38]

Chandler Brooks, associate Professor of Physiology at the Johns Hopkins Hospital Medical School came to Dunedin in September 1945 to work with Eccles for a year on the spinal cord. While in Dunedin, Brooks and Eccles published eight joint papers. Some of the titles of these joint papers were as follows: *An electrical hypothesis of central inhibition,* (1947a)[39]; *Electrical Investigation of the Monosynaptic pathway through the spinal cord,* (1947b)[40]; and *Synaptic Potentials of Inhibited Motoneurones,* (1948)[41]. Chandler and Nelle Brooks remained close family friends and Jack and Rene later visited them in America.

Jack and Rene's eldest daughter, Rosamond, began university in 1945, finishing her degree in 1948 and later starting her Ph.D. in Cambridge. (She later worked in her father's department at the Australian National University).

Visit to America 1946:

Now that World War II was over, people were free to move around the world and passenger aircraft would soon be operating between countries. After receiving an invitation from the New York Academy of Sciences to participate in an international neurophysiological conference, Jack Eccles left for America on his first visit on 17 January 1946. To get there he had many strange methods of travel. He left Wellington on a slow American Navy cargo boat to Hawaii. The *Sword Knot,* of 2,300 tons, was delayed by head winds but he managed to make the connection in Hawaii with a flying boat - an old China Clipper - an enormous aircraft with a fine dining room, bathing facilities and sleeping bunks. When he arrived at San Francisco, he had to use the name of the Rockefeller Foundation and the New York Academy of Sciences to get his ticket for the final leg of the journey. Flying on a TWA DC3, he arrived in New York on the very eve of the Congress and only half an hour late for dinner! The Rockefeller Foundation covered the cost of the whole trip.

On his return journey, in February 1946, he had a seat on a DC4 which had stopovers at Honolulu, Canton Island, and Fiji and then finally Melbourne. It was not a passenger flight, but was a delivery flight for a new aircraft for Trans Australian Airlines, a newly formed Australian domestic airline. It carried 20 passengers by special arrangements, with a steward who was recruited in San Francisco and who arranged for the provisioning. The pilot and navigator had not flown that route before and they used a National Geographic map of the Pacific Ocean that they had bought in San Francisco! This was the first flight by a commercial company across the Pacific.

Decision to leave New Zealand:

Jack Eccles had come to the realisation that if he wanted to continue on the worldwide scientific scene, he needed to have more time for research in neurophysiology. The heavy teaching load in Dunedin was hampering his progress. In the late 1940's, he began to hear of plans in Australia to begin a Research University in Canberra. When he was offered the Chair of Physiology in 1950, he had no hesitation in accepting it. The person who replaced him in Dunedin was Archie McIntyre who had been a senior lecturer since 1949. Another lecturer was Marianne Fillenz, who had been one of his earliest pupils in Dunedin. In 1950, she left for Oxford University to continue her career in Physiology with the encouragement of Professor Eccles. The most interesting associate Eccles had in Dunedin was Ainsley Iggo, who had specialised in Agricultural Science. He then did the new course at Otago University initiated by Eccles – B.Sc. Physiology. Ainsley went on to become a professor of Veterinary Medicine in Edinburgh and became a Fellow of the Royal Society in 1978.

---

[38] My Scientific Odyssey, 1977:6.
[39] Nature, 1947, 159:760-64.
[40] Journal of Neurophysiology, 10:251-74.
[41] Journal of Neurophysiology, 11: 417-30.

Research continues:

Although he had accepted a position in Canberra in 1950, Jack Eccles was not able to make the move there until 1952, and, in the meantime, his research continued into transmission across synapses: By 1949 the great weight of the evidence had caused him to repudiate electrical transmission at all the peripheral synapses, but with synapses in the central nervous system the problem was still open. He continued to espouse the electrical theory both for excitation and inhibition in this area.

He was always at the forefront of the introduction of new techniques for research:

> While in Dunedin, Eccles decided to test the electrical hypothesis regarding transmission across synapses in the central nervous system. To do this he introduced in 1951 a new technique of intracellular recording from motoneurones in the spinal cord. This involved the insertion of a glass micro-electrode, 0.5 to 1 micrometers external diameter, into a motoneurone of the spinal cord, making it possible to record action potentials of a single motor nerve cell.[42]

As Eccles had become a disciple of Karl Popper, he was quite happy to disprove his own theories. He would postulate a hypothesis that would even attract the possibility of falsification. One memorable experiment to testing his hypothesis about electrical or chemical transmission took place in August 1951 in Dunedin as he describes it:

> Thus it was a clear test. If the quadriceps volley caused the trace to go up it was electrical, if down it was chemical. It went down... We were momentarily stunned, as well we might be after a long day's experiment, enhanced by some extraordinary obstetrical complications. The wife of one of my two associates (Jack Coombs) was delivered of a baby girl by the other (Lawrence Brock), I meanwhile tending the experiment. It was by then in the early hours of the morning. But on recovering from the shock (the physiological, not the obstetrical) the decision was made. Inhibitory synaptic action was chemically mediated and it was evident that the mirror image response, excitatory synaptic action, was also chemical.[43]

To carry out these experiments, Jack Coombs designed and built several recording units, known as Electronic Stimulating and Recording Units (ESRU). These units allowed for very detailed records to be collected during the course of the work. They were "four magnificent electrical stimulating and recording units designed by Jack Coombs and built in New Zealand. At that time and for many years to come – in fact until the transistor era – they were the best general research instruments for electrophysiology in the world".[44]

In 1952, Eccles, Brock and Coombs published several papers. One of them was entitled, *The Nature of the Monosynaptic Excitatory and Inhibitory Processes in the Spinal Cord.* This paper, based on work done at the Medical School in the Otago University, showed that, even in the spinal cord, the synaptic transmission was a chemical process.

Overseas travel, 1951 to 1952:

In the transition period from Dunedin to Canberra, Eccles travelled to the United States in 1951-52 to meet with colleagues including Chandler Brooks and then to England where Hodgkin, Huxley, and Keynes in Cambridge and Fatt and Katz in London were all leading the world in neurobiology.

In 1951, at a meeting of the Physiological Society, he suddenly revealed that he had proved himself wrong in his previous theory; that it was chemical rather than electrical transmissions between the synapses that causes the neurons to be fired. The physiologists were astounded because Eccles had previously, strongly and bitterly, debated his views with Dale who had always held that it was a chemical transmission. Eccles was able to do this without too much loss of face because of the influence of his friend, Karl Popper, who held that, in science, it is acceptable to disprove your own theories. Dale described Eccles' talk as a remarkable conversion indeed! "One is reminded almost

---

[42] Australian Science in the Making, 1988:303.
[43] Closing Address of the Sir Henry Dale Centennial Symposium, 1975, quoted in Australian Science in the making, 1988:304.
[44] Australian Science in the Making, 1988:304.

inevitably, of Saul on his way to Damascus, when the light shone and the scales fell from his eyes." Some years later Eccles realised that he had capitulated too soon. By 1964, he found that, while many of the synapses he worked on were indeed chemical, many electrical synapses had also become known and he included two chapters on electrical transmission in his book *Physiology of Synapses* both excitatory and inhibitory.[45]

In 1952, Jack Eccles spent five months in residence at his old college, Magdalen College, preparing and delivering the Waynflete Lectures, named after William of Waynflete, Bishop of Winchester, who founded the College in 1458. The title of the lectures and the subsequent book was *The Neurophysiological Basis of Mind, Principles of Neurophysiology.*

Eccles meets Sherrington for the last time:

On 23 February 1952, Jack went once more to see his old mentor, Sir Charles Sherrington. He was rather apprehensive at the thought of seeing him so deteriorated. However, all went well. They talked about Sherrington's book, *Man on his Nature*, and the great problem of the design and functioning of the brain that gave it the transcendent property of being in liaison with mind and the mental world. At that last meeting, Sherrington said, "For me, Jack, the only reality now is the human soul", something that he was always to remember.

Nine days later, on 4 March 1952, the tolling of the Magdalen bell made Jack suspect the worst - their greatest Fellow had died.

Eccles family in front of the Dunedin house. Back row: Judith and Frances. Middle row: Jack, Peter, William and Rene. Front row: John, Rose, Richard, Alice and Mary.

[45] Facing Reality, 1970:106

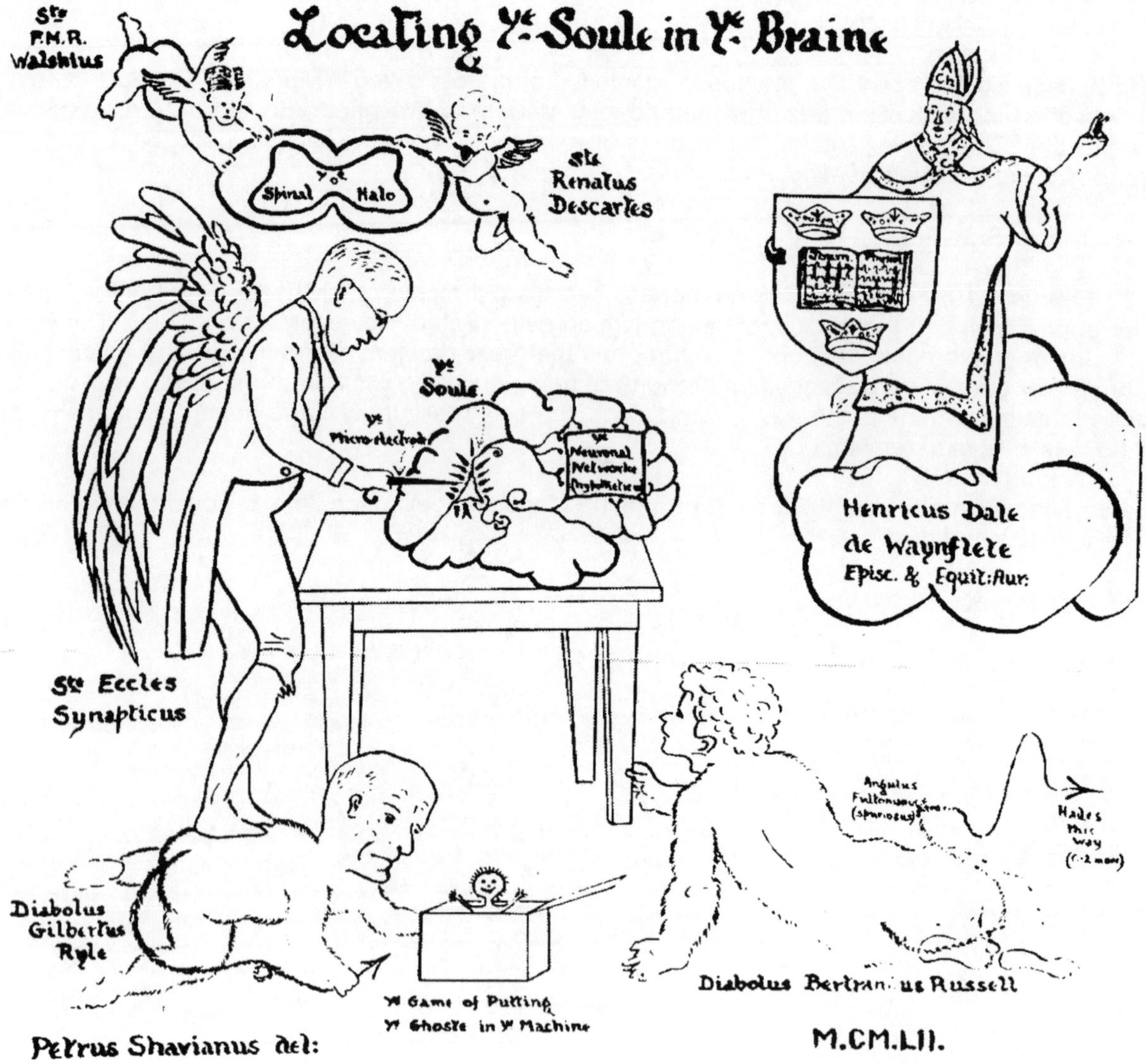

Locating the Soul in the Brain. The four main characters are: 1, Jack Eccles, the angel cutting the brain to find the soul; 2, Gilbert Ryle, depicted as a devil, was a philosopher who wrote 'The Concept of Mind' (1949). He was an opponent of Cartesian dualism which was caricatured as the 'ghost in the box'; 3, Sir Henry Dale, depicted as William of Waynflete, founder of Magdalen College, bishop and knight; 4, Bertrand Russell, depicted as a devil, was a philosopher and atheist. The two angels at the top represent F.M. Walsh and René Descartes whose Dualist Theory of distinct body and soul was often quoted by Jack Eccles. The cartoon, by Peter Shaw, was published in 1952 at the time that Jack Eccles was delivering the Waynflete Lectures.

Chapter 5:

Canberra, 1952 to 1966:

The aim of the new university in Canberra, later to be called the Australian National University (ANU), was to attract world-class people in their fields and to establish a research university. Prime Ministers Chifley, Curtin and Menzies all gave generous support. As it was intended that the ANU would specialise in post-doctoral research, it was expected that it would attract both Australian and Overseas students of a high calibre.

The University was pleased that Jack Eccles had accepted a position at the University. The Council of the ANU considered his acceptance as a "brilliant catch" considering he could have accepted positions in England and the United States. It was Sir Howard Florey's decision to tempt Eccles across the Tasman to Canberra as he considered Eccles as "one of the best living neurophysiologists". Eccles himself thought that the ANU would give him "exceptionally favourable opportunities for pursuing his research interests and a release from his current burden of teaching at Otago".[46]

Canberra at that stage was not much bigger than a country town of 25,000 people with sheep grazing around the parliamentary buildings. The land for the university was only a large grassy field dotted with gum trees and with two army huts for administrators, but there was good financial support by the Australian Government and a vision of the buildings that would soon be constructed.

The Eccles family settled into their new house in Canberra in September 1952. It was surrounded by 2½ acres (one hectare) of virgin soil. There was plenty of room for fruit trees and a large vegetable garden as well as a shrub and flower garden, a tennis court and a small swimming pool.

John Curtin School of Medical Research:

Before the war, there was no national medical research institute, although there were several privately endowed institutes such as Kanematsu and the Kolling Institutes in Sydney and the Walter and Eliza Hall Institute in Melbourne. However, towards the end of the Second World War, Howard Florey developed the concept of a National Medical Research Institute for Australia, similar to that in England. He mooted the idea when he visited Australia to advise the government on the production of penicillin. This dream of a National Institute was realised with the creation of the John Curtin School of Medical Research, as an integral part of the Australian National University in Canberra.

The first professor at the John Curtin School of Medical Research was Professor Hugh Ennor in Biochemistry, followed by Professor Frank Fenner in Microbiology and Dr Adrien Albert in Medical Chemistry. The fourth one was Professor John Eccles in the Chair of Physiology. As one of the founding professors of the ANU, Eccles helped establish the John Curtin School of Medical Research and saw the research he did there as the highlight of his career. He also saw it all as a challenge, a new venture and was enthusiastic about it.

When Eccles moved to Canberra, he took Arthur Chapman, his technician, and Jack Coombs with him as he needed their expert advice on the equipment he used. Jack Coombs was a physicist and electronic instrument designer and the equipment he designed "was critical to Eccles' research during his fourteen years in Canberra".[47]

---

[46] The Making of the Australian National University, 1996:60.
[47] The Making of the Australian National University, 1996:91.

Photograph by courtesy of John Curtin School of Medical Research.

Professors Eccles, Ennor, Albert, Fenner and Bunker.

Photograph by courtesy of John Curtin School of Medical Research.

Staff of the John Curtin School of Medical Research.

Eccles was always able to attract expert technical assistance to design, install and maintain his laboratory equipment. Over the years, these people included Coombs, Chapman and, from 1958, Lionel Davies, who joined the Department as the Head Technician. Lionel came from Queensland where he had gained a Diploma in Mechanical and Electrical engineering from the University of Queensland. While Jack Eccles was in charge of the whole department, he left the technical side, including the design of new instruments, in Lionel's capable hands. (Lionel worked in Canberra for many years and then accompanied Eccles to America where he set up the necessary equipment to facilitate Eccles' continuing research in Chicago.)

In the early 1950's, setting up the new Department was a real challenge for both Coombs and Eccles. As with the rest of the University, the buildings for the Medical School were far from ideal. The first laboratory was in a temporary building, but even this was not available until February 1953, with a larger building planned for the future. Initially, Eccles only had a room for himself and one for his secretary and he filled in his time working on the final drafts of his book, *The Neurophysiological Basis of Mind,* which was to gain him an international reputation in the Mind/Brain field of research and became a set text in medical as well as veterinary courses. It "described his investigations of the process by which nerve impulses travel along nerve fibres and are now transmitted and now blocked at the synapses".[48]

Jack Eccles always had a great capacity for work and carried out long experiments at the labs two or three times a week. Dr David Curtis explained how the work was undertaken:

> When an experiment was arranged to take place, a group would start work at 7.00 to 7.30 in the morning to prepare and anaesthetize the animal and prepare equipment. By 5.00 or 6.00 p.m. everything was ready for the cat to be moved into one of the laboratories, which were shielded against electrical interference. They would then work all night, going home briefly for a shower and breakfast and returning to work on the cat throughout the next day and perhaps the day after that.[49]

One day as the Eccles children were preparing for school, their father arrived home from work and said, "Guess what I discovered last night?" They had no idea. "Well I discovered how tetanus affects the nervous system." The family was impressed by his discovery, but also impressed by the fact that he had worked all the day before and through the night without sleep. Of course, tetanus injections had been available as a treatment before this but it was not known how the disease affected the body. Subsequently, they were all lined up for a tetanus injection, except for young John who climbed up a gum tree and refused to get down. As a result of his findings several papers were published on the subject including one called, *Mode of action of tetanus toxin* with his colleagues, V.B.Brooks and David Curtis.[50]

Professor Eccles' Department soon took on an international appearance. Researchers from many countries came to work with him, among them his daughter Rosamond, A. Lundberg, David Curtis, Paul Fatt, Per Andersen and several Japanese scholars, the most notable being Masao Ito. An example of the work that was done in this era, is described by Drs Curtis and Andersen: "With his daughter R.M. Eccles and A. Lundberg, Eccles initiated in 1956 a series of papers on the neuronal organization within the lumbar spinal cord using intracellular recording of postsynaptic potentials".[51]

During this time, many scientific papers were written and four books were published including *The Neurophysiological Basis of Mind* in 1953; *Physiology of Nerve Cells* in 1957; *Physiology of Synapses* in 1964; *Cerebellum as a Neuronal Machine* in 1967.

---

[48] Ian Howells, Catholic Leader, May 1997.
[49] The Making of the Australian National University, 1996:92.
[50] Nature, 1955, 175:120-121.
[51] Historical Records of Australian Science, Vol 13, No 4:446

Temporary Laboratories 1952 to 1957.

Photograph by courtesy of John Curtin School of Medical Research.

Professors John Eccles, Adrian Albert, Frank Fenner and Hugh Ennor study the plans for the new building.

Looking back at this time it was noted:

> Eccles made great discoveries concerning the workings of the central nervous system at its synaptic level. During the 1950's and 1960's, he created a laboratory in Canberra that was the centre of the world in terms of central nervous system research. His proteges have led science in their own nations, including the chief scientist of Japan, Masao Ito, who was trained by Eccles; the leading scientist in the former USSR, Kostyuk; and the most famous scientist in Norway, Per Anderson. His ideas and experimental influence live on in a dramatic way. These great men, who still work on the areas to which Eccles introduced them, are part of his legacy to the world.[52]

Masao Ito remembers the pace of work in the department:

> Every Friday evening, a seminar was held on the work done in the department – and with the approach of the weekend, the atmosphere became tense: we were exhausted from the hectic work and vigorous discussions. --- But each Monday the Prof appeared in the lab, marvellously refreshed. I believe the secret to his rejuvenation was the time that he spent in his big garden and in simple study at home.[53]

It was not until 1957 that the new John Curtin School of Medical Research Building was finished and the scientists were able to vacate their temporary accommodation. Eccles wrote, "At last the grandiose new building was completed. We had done so well in the temporary hut that I was somewhat overawed by the new magnificence and the greatly extended facilities".[54] Since Oxford days, when he had to share with Denny-Brown, Granit, Olmsted and Marcu, he had seen the need for individual space and planned accordingly.

In Canberra, his research has been described as his golden era when he could concentrate without the interruptions of undergraduate students and course work to be prepared. He said of this time:

> My interests were concentrated on neuroscience, which provided the scientific challenge of my life, particularly the 14 years at the Australian National University at Canberra (1952 - 1966). I did not engage in philosophical adventures on dualist-interaction, but I accepted invitations to lecture and write on the mind-brain problem, which had remained much as I left it in 1952. Also, experimentally, I left the spinal cord and brain stem for the cerebral cortex, the hippocampus. Then in 1963, I made a great departure to the cerebellum, from 1963 to 1976, wryly remarking that in the cerebellum I could study neuroscience without being involved in the possibility of disturbance by conscious phenomena! [55]

An article in the Sydney Morning Herald stated that, between 1963 and 1967:

> Sir John had been concentrating on the cerebellum, which is connected with the co-ordination of muscular movement. Instead of studying single cells, he is now looking at the inter-relationship, the "hook-up" or "wiring-pattern" of cells in assemblages. It is from this kind of pure research that major advances in medicine come. The application of Sir John's discoveries about the nature of the brain to the treatment of mental and nervous disorders like schizophrenia is only just beginning. But the impact of his work will be felt not only in practical fields of medicine. It has also contributed to a fast-changing view of the way man organises the information which his senses bring to him.[56]

---

[52] Medical Oberver, 11 July 1997.
[53] Nature, Vol 387, 12 June 1997.
[54] My Scientific Odyssey, 1977:11.
[55] ibid
[56] Science and You, 20 February 1967.

***Above left:* Jack Eccles was proud of his residence. *Above Right:* Dr K. Frank with the family on a mountaineering expedition.**

***Below:* Jack and Rene Eccles, Richard, Frances, Mary, Judith and Dr K. Frank on the way to the South Coast.**

During his time in Canberra, Sir John travelled widely overseas and, more locally, to Papua New Guinea in 1957. He gave slide evenings after each of his trips, first making sure that no one in the audience had been to the place (he didn't want to be upstaged by an expert). At one such talk, he had fascinating insights into the Kuru Brain Disease in Papua New Guinea caused by the adults eating the brains of those who were already infected with the disease. He told us that the people in PNG needed help in the schools and in new systems of agriculture. Two of his children, Mary and John, spent years in Papua New Guinea and their interest was first sparked by this slide evening.

Also in 1957, he published *Physiology of Nerve Cells* which was based on a series of Herter lectures at the Johns Hopkins School of Medicine. The Quarterly Journal of Experimental Physiology said of this book, "it provides the clearest and most complete account of the variety of devices developed by the nervous system for the transmission of impulses, and provides a solid basis on which to build schemes of neural organization".

Another book, *Physiology of Synapses* was published in 1964 by which time Jack Eccles realised that he had made too hasty a retraction of his theory on electrical transmission back in 1951 when he had faced Dale across the room at the Physiological Meeting. At that time, he had decided that the impulses were chemically transmitted as Dale and his team had believed. But now, as already mentioned, it turned out that "many types of synapses I had worked on are certainly chemical, but now many electrical synapses are known".[57] So, both sides of the debate were partly right and partly wrong.

Home life:

The overseas scholars often visited the Eccles family for tennis parties at the weekend. Rene would cook her special scones with delicious homemade raspberry jam and cream. The afternoon tea would usually be spread out on the lawn beside the tree belt and tennis court. David Curtis and his wife, Laurie, and their children, Christopher and Belinda, would be frequent visitors in those days. Jack Eccles was trying to duplicate the Sydney days when he, Kuffler and Katz did the same on the tennis court in Mosman.

The big garden mentioned by Masao Ito extended over two acres - vegetables grown in the virgin soil were enormous - carrots, potatoes, rhubarb and spinach with prize-winning sized leaves. All members of the family were engaged in planting, weeding and rotary hoeing. Along the fence, down the bottom of the vegetable garden, were boysenberries and raspberries, and in the orchard peaches, apricots, plums and nectarines flourished. Excess fruit was collected and bottled or turned into jam. Jack Eccles occasionally appeared at the back door with wheelbarrow filled with tomatoes which he would cut up for bottling or cook for breakfast. It was marvellous the work that Jack put into the garden. He sprayed and pruned the fruit trees and, when the fruit was ripe, he revelled in the fresh produce, although, occasionally, he found the garden time consuming and somewhat of a burden.

While he worked in the garden, Jack would think over the problems in his work and then turn again to the labs refreshed with new ideas. There were many keen gardeners among the people of Canberra and amongst the academics in particular. It was interesting to hear the professors from the university talking. As well as discussing their own fields of endeavour, the conversations ranged from university politics to world affairs, but often they would talk about their gardens. Professors Ennor, Fenner, Oliphant and others grew much of their own produce in those days.

The folk dancing nights were on again as they had been in Dunedin and people from all walks of life came and enjoyed the evenings. Neighbours, our school friends, Professor Mark Oliphant, his wife Rosa and their daughter, Viv, Professor and Mrs Passmore and their daughters, Helen and Diana, were regulars. We all had lots of exercise and many a laugh around the pool on those Friday nights.

Among visitors to the house were the Prime Minister, Sir Robert Menzies and Dame Patti who enjoyed a very convivial afternoon. Dame Patti was a wonderful, warm person who got on well with Rene. Jack said of Menzies, "He was one of the really great men I have been privileged to know --- Menzies advised

---

[57] Facing Reality, 1970:106.

that the first permanent building to be erected [at the ANU] would be the residential college and faculty club, called University House. He rightly sensed that a university is a community of scholars and that there had to be good facilities for cross-cultural meeting in a faculty club".[58]

The South Coast of New South Wales:

Through the years in Canberra, the Eccles family would holiday down the South Coast at Kioloa. There, for a few pounds, one could rent a cottage near the beach. There were miles of unspoiled beaches in those days, Merry Beach, Pretty Beach, Pebbly Cove, all of which Jack was partly instrumental in getting declared national parks. There were no lifesavers in the area and the family occasionally saw large sharks swimming along in the curve of the waves. There was the day that Jack put his life on the line when he swam out to rescue two of his children caught in a swirling rip.

Jack Eccles never lost his love of nature, which he had gained from his father in early childhood. From Sherrington, he had learnt to be cultured and to have an interest in the arts as well as in the scientific world. In fact, with him it was hard to see the blurring between the scientific and the natural world. His wonder at the brain and the mind dovetailed with his wonder at the natural world - he did his most concentrated work lying on a thin mattress sun baking beside the flowers near the swimming pool. At night he used a green tennis shade to keep the light out of his eyes as he studied at his desk. However his deftness at the slide rule left his children amazed. If they had a problem in maths he could quickly whiz up the answers but had no idea of all the intermediate steps.

Australian Academy of Science:

In 1951, Professor Mark Oliphant and D.F. Martyn, a scientist from the CSIRO, mooted the notion of an Australian Academy of Science based on the Royal Society in London. It was formally inaugurated in 1954 and opened by the Queen. It was the second time in history that a British Monarch had opened a scientific society, the first being the Royal Society of London by Charles II in 1660. Jack Eccles was a Foundation Fellow of the Academy.

Because of Oliphant's involvement, he became the First President from 1954 to 1957. Jack Eccles was the second President from 1957 to 1961. The actual Academy building was finished in 1959. A striking symbol of Australian Science, it was an enormous copper-covered dome, a section of a sphere, broken only by arches rising from some 16 bases immersed in a circular moat. The scientists were overjoyed at their symbol and the local population was agog at the sight. It promptly became part of the tourist bus circuit. Over the years, the Academy of Science has taken an increasing interest in science taught in school and has developed programs to teach science even at the primary level. So

[58] My Scientific Odyssey, 1977:10.

Sir John Eccles being knighted by Governor-General, Sir William Slim.

Below: Cartoon by Pam McFarlane.

the vision those pioneer scientists, like Eccles and Oliphant, had of developing a scientific centre where scientists could congregate proved to be a great success. Excellence could be recognized by the award of Fellow of the Australian Academy of Science, F.A.A. "Awards could be of immense value to institutions as proof of achievement."[59]

Knighthood 1958:

In 1958, Eccles took his birth name, John, rather than Jack, when he was honoured in the Queen's birthday list, becoming Knight Bachelor. The day of the investiture ceremony was cold and windy. The families assembled outside Government House and were greeted by the Governor General and his wife. Sir William Slim, the Governor General, looked stunning in his military uniform and knighted Eccles and the others with all the pomp and ceremony he could muster.

Afterwards, everyone was ushered outside so the chairs could be put away for the reception. Mary remembers, "It was cold and our hair was blowing in the wind. No one looked their best and who should be there to congratulate the new Sir John Eccles, but Robert Menzies, the Prime Minister. Of course, we were introduced and tried to make conversation through chattering teeth and wind-chaffed faces to this famous man with his bristling bushy eyebrows and his kind face. It was a momentous occasion. Menzies made a comment that he liked the idea of knighthoods as the wives shared the award taking the title Lady. In this case it was Lady Irene Eccles".

False Alarm:

One day late in 1960, word filtered back to Australia that there was an Australian on the short list for the Nobel Prize in Medicine. That evening, it was announced on the ABC that Sir John Eccles was the

---

[59] The Making of the Australian National University, 1996:253.

Photograph by courtesy of John Curtin School of Medical Research.

Sir John Eccles in his laboratory.

recipient along with a Professor McGoon in America. Professor Eccles was at the labs when journalists contacted him and he and his colleagues were delighted with the news. He immediately rang home and said, "get the champagne out and invite the neighbours over". Everyone was very happy and he arrived home overjoyed at the news. However, a short time later, the ABC rang and apologized for their mistake. The next news broadcast announced that it was Sir Macfarlane Burnet who had won.

Naturally Eccles was devastated and the party was over before it began. He was half way through an experiment at the time of the false alarm, so, rather crestfallen, he left the gathering and drove back to the labs to immerse himself once more in his work. It shows the calibre of the man that he could do this. On his arrival back at the labs, his colleagues melted away, not willing to greet him but Del Doherty, in an adjoining lab, came forward and shook his hand. "Sorry about the news," she said, "but I hope I'm here to congratulate you when you really do win." Three years later when this occurred, Eccles wrote her a special note because she had raised his morale when he needed it.

Honorary Doctorate of Science at Cambridge:

In 1960, the University of Cambridge awarded Honorary Doctorates on the occasion of the Tercentenary of the Royal Society. One of the recipients was Sir John Eccles.

The Citation at the presentation of his Honorary Degree in Cambridge reads:

> Among the Australians who have been sent to the University of Oxford by the forethought and munificence of Cecil Rhodes this man holds an eminent place. For having formerly as a pupil obtained a brilliant teacher in Sir Charles Sherrington, he was soon able as a colleague to repay his teaching through his skill in new techniques of research which could be applied to the investigation of the animal body. He himself has examined the minute cells of the spinal cord in order to acquire intimate knowledge of the nature and functioning of the nerves first discovered by Herophilus. Aristotle surmised that the impulses arising from sensation made on our minds imprints like those of signet-rings, and that men remember and recall things when one impulse follows another by natural connection. This man is perhaps on the way not merely to surmising but to discovering at last traces of such imprints in the brain.
>
> So much for scientific research, which he pursues with energy and success. But I must not pass over his home life in silence. How so? If the ancient Romans gave special rights to those citizens who had three children, what rights should this man be given who has three times three?
>
> I present to you Sir John Carew Eccles, President of the Australian Academy of Science.

Continuing interest in his children:

As the Citation commented above, Sir John was a family man who supported his children when he could. This is no more illustrated than when, in 1961, he supported his son, Peter, who had some trouble in Albuquerque where he was doing cosmic ray research. Concerned that it appeared Peter was not getting the credit for his work and that others were apparently taking his research data, he did his utmost to help. He wrote, "Let me tell you Peter that I feel just as mad about all this as if it were actually done to me and certainly if something like this happened in my Department there would be pretty tough action in milliseconds". To a colleague in America he wrote "I am scheduled to give a lecture at Harvard on the 29 September (1961) but if I thought I could be of assistance in a crisis, I would certainly go to the expense and inconvenience of an earlier visit to Albuquerque". Later when things cleared up, he wrote to Peter, "I am delighted to hear of your contacts at the Mexico meeting. I am sure that there are lots of opportunities for you with other people in America". [60]

The Nobel Prize, 1963:

In October 1963, Sir John Eccles received the Nobel Prize with Alan Hodgkin and Andrew Huxley for Medicine or Physiology, "for discoveries concerning the ionic mechanism in the excitation and inhibition of the peripheral and central membranous sections of nerve cells".[61]

Alfred Nobel (1833-1896) willed most of his vast fortune, made from his invention of dynamite, to establish a series of prizes to be awarded each year in the fields of physics, chemistry, medicine, literature and peace, to men and women who had been the most outstanding in each field. The candidates were selected by various Swedish academics for all the prizes except the Peace Prize, which was decided by the Norwegian Government. To win a Nobel Prize is regarded as among the highest honour in the world and it has worldwide recognition.

The awarding of two Nobel Prizes in the 1960's, Macfarlane Burnet in 1960 and John Eccles in 1963, for research carried out in Australian Research Institutes was seen as a climax of the post-war effort.

> The environment for research in the medical sciences in Australia in the 1960s had changed dramatically from that which existed before the Second World War, a just reward for those who had struggled for so long to reach their goal of national independence. Australian scientists responded when the stage was set for them to work in their own country under conditions that enabled them to compete in the international sphere of medical research. [62]

Sir John wrote: "I received the Nobel award (1963) for the ionic mechanisms of synapses, and my Nobel Lecture was on the ionic mechanisms of postsynaptic inhibition." He was not in Australia at the time it was announced, but in Venice with Rene on one of their many overseas trips. They were out on a gondola when the news arrived at the Venetian government office. A messenger went out after them with the message and the gondoliers made much of him when its importance spread. He was photographed in St Mark's Square with the gondoliers, on a Venetian bridge and feeding the pigeons.

The three Nobel Laureates arrived at Stockholm Airport for the official ceremony at the same time in December 1963. There was Alan Hodgkin and his family, Andrew Huxley and his family, but Jack and Rene Eccles were alone. None of their nine children was able to be there. King Gustav of Sweden made the presentation. During the celebrations, Sir John Eccles sat next to a Swedish princess at the Nobel dinner and later danced with her sister. It was a great occasion and he enjoyed it immensely. He joked with his fellow laureates, after surveying the elderly winning scientists, in a booming voice, "We physiologists are obviously the healthiest of the lot".

---

[60] Personal letters in Dr Peter Eccles' files.
[61] The Nobel Prize, 1963.
[62] Australian Science in the Making, 1988:304.

Above: Sir John and Lady Irene Eccles with the gondoliers in Venice after being told that he had won the Nobel Prize.

Below: In the presence of other members of the Swedish Royal Family, King Gustaf Adolf presents Sir John Eccles with his Nobel Prize.

Photograph by courtesy of News Limited.

His latest work, called *The Physiology of Synapses,* "was published just in time to be on display in Stockholm at the time of the prize festivities in December 1963. The publishers made an unprecedented effort in speed of publication in order to effect this felicitous timing".[63] This book described the work of Eccles and his colleagues in Canberra on the investigations of the patterns of the organisation that exist in the spinal cord. This work used the techniques of recording from individual cells in the spinal cord.[64]

An article in the Sydney Morning Herald explained the significance of Sir John's work in layman's terms:

> One of the great modern advances in the understanding of the function of the human brain was made by an Australian Scientist, Sir John Eccles. For nearly 20 years he concentrated on a single aspect of brain performance – the way in which an impulse or "message" is transmitted from one nerve cell to the next.
>
> It was Sir John's achievement to record the electric potential of single nerve cells by means of incredibly thin glass needles connected to electronic apparatus. Sir John based his theories on the earlier work of two British scientists Professor Alan Lloyd Hodgkin and Professor A.F.Huxley. They were primarily concerned with the transmission of impulses along the nerves rather than between one cell and another. With the aid of tiny electrodes inserted in the spinal cord, Sir John was able to show that there are in fact two types of nerve cell. One "fires" messages at a great rate through a process known technically as excitation. Another kind of nerve cell, he found, acts in the reverse direction, tending to damp down or "dull" the transmission of messages by a process known as inhibition. He was the first person to demonstrate active inhibition, although inhibition had been a theory for many years. It now seems that there are as many "dulling" cells as there are "firing" cells and the proportion plays a vital part in the healthy functioning of the brain.[65]

The scientists who chose Sir John as a Nobel Laureate found that it was not a lucky chance discovery that gave him his scientific results but the result of "sedulous, systematic labours":

> Nor was it an incident (sic) that Eccles picked the motor cell as a target for his micro-electrode, which has a point measuring less than one-thousandth of a millimetre across. The motor cell in the spinal cord is the last cell of the reflex arc from which the nerve fibre to the muscles emanates. It is the motor cells that set signals going to the muscles. The signals entering the spinal cord from the sensory cells react on the motor cell. That the peripheral nerve fibre sends its signals as all-or-none impulses was already known. ....
>
> He knew the nerves which excite and the nerves which inhibit a certain group of motor cells. Once Eccles got up steam with his technique, the answers came in rapid succession. It soon became clear that the motor cell when active has an all-or-none impulse, which in all essentials resembles the impulse in the peripheral nerve. The cell has at rest a membrane potential corresponding to that of the peripheral nerve. The membrane starts a regenerative activity when its potential is reduced to a critical threshold value. During the impulse the membrane undergoes a resistance or conductance change.
>
> Eccles immediately related his discoveries to the ionic theory formulated by the Cambridge school for the squid nerve. Hodgkin and Huxley published their voltage-clamp analysis in 1952, the same year that Eccles began his micro-electrode research. ....
>
> We see here how two important areas of neurophysiology have been radically altered by two research teams. In each case a dimly understood and complex sequence of events was explained with reference to some specific changes of permeability. Future research is bound to subject these permeability changes to further analysis. The research which merited this year's

---

[63] My Scientific Odyssey, 1977:13.
[64] Australian Science in the Making, 1988:304.
[65] Science and you, Sydney Morning Herald, 20 February 1967:3

Nobel award is extremely fundamental and highly essential to an understanding of how the nervous system functions. [66]

The whole family together for the last time. Jack, Rene, Rose, Peter, Alice, William, Mary, John, Judith, Frances and Richard. January 1964.

Australian of the Year, 1963:

Sir John Eccles was the Australian of the Year for 1963. The Award was announced in January 1964 by Sir Norman Martin, chairman of the Australia Day Council.[67] It was a busy time as he was much in demand by reporters, having just won the Nobel Prize. There was also a wedding in the family in January 1964 and the whole family was present. It had been the first time for many years that there had been a family gathering and, as it turned out, it was the last time as well.

The Pontifical Academy:

Sir John Eccles was the first Australian to be elected as an Academician of the Pontifical Academy in Rome. In 1964, he was responsible for a Study Week with the Pontifical Academy of Science at the Vatican from 28 September to 4 October. The purpose of the week was to relate Psychology to the Neurosciences. The title of the symposium was *Brain and Conscious Experience*. In the preface of the subsequent publication he wrote:

> The symposium was held in the Academy building in the most delectable spot in the Vatican Gardens, being built in 1561 as a Casino for Pius VI. The room in which we met had still the lovely tiled floor and the frescoed walls and ceiling of that period, and outside there was always the sound of the fountains in the marble paved court. Then we all lived (most of us with our wives who were also guests of the Academy) in the stately Hotel Reale, where we were magnificently entertained. We returned there between the sessions of each day for a sumptuous lunch with time even for a brief siesta before the late afternoon session. And in the evenings, there were the pleasant social occasions of dinner and gatherings thereafter. From a rather large experience of symposia in the last decade I am convinced that the most fruitful have been the most enjoyable, because it is under these conditions that the human spirit (if I may use this phrase) operates most creatively in debate.[68]

His Holiness, Pope Paul VI, who granted a special audience to the scientists, welcomed them to the symposium saying, "We have desired to bring you personally Our greetings and Our thanks and to express anew the interest with which We follow the development and the progress of your scientific activities. --- May We be permitted simply to underline in a word its importance, and to bring out its relationship with those domains in which the essential part of Our own activity is exercised. We refer to the moral and religious sciences".[69]

---

[66] The Nobel Prize, Annual Edition 1963:30-31.
[67] The Advertiser, 16 January 1964.
[68] Brain and Conscious Experience, 1966: viii.
[69] Brain and Conscious Experience, 1966: xix.

Chapter 6:

America, 1966 to 1975:

Retirement Age:

In 1966, Eccles was 63 and, knowing that the compulsory retiring age at the ANU was 65, he tried to get this extended to 68 but the administrators prevented this. He said, "I already knew the very impoverished conditions that would be my lot after 65 – half salary, renewable year by year, one laboratory and with almost no support for staff or assistants."

The issue was much more widely advertised than just at the university. Across the country, his views were noted. He gave a lecture to the Australian Association of Gerontology in February 1966, where he said:

> Today about two-thirds of men over 65 in Australia are no longer at work. Many of these have been retired when they were still very valuable with their skills, wisdom and knowledge. No doubt many should be retired. But the aim of society should be to introduce more flexibility in the whole question. In fact my plea is that the effective working contributions of the aged be maintained for as long as possible. Many people would find their own family life replenished if the grandparents, though not under feet, were near at hand. There are special bonds of sympathy between children and their grandparents' generation.[70]

Cartoon by courtesy of Herald and Weekly Times Ltd.

That same day in The Canberra Times there was an article headlined:

> U.S. position for Sir John Eccles.
>
> In Chicago, Sir John, who specialises in the function of the nerves, plans to continue to map the structure, function and nerve patterns of the cerebellum, the part of the brain responsible for muscular co-ordination. He said he would be working in the Institute of Biomedical Research in Chicago. "I leave Canberra with regrets but with great anticipation".

Even the Canberra Times Editorial took up the subject:

> It would be an over-simplification to say that Sir John Eccles is being driven abroad solely because if he were to stay at the Australian National University his research would be cut short within a year or two by retirement provisions. The offer from the Institute of Biomedical Research in Chicago would have been a vastly attractive one at any stage of his career. But in accepting it he has drawn attention to an important consideration. In Chicago there is no fixed retirement age, he will be able to continue his work indefinitely. At the ANU he would find himself only of limited use after the age of 65: he would no longer be chief of a research team. But most people will not be concerned with the natural complexities of a scientist's motives. What will concern them, and rightly, is the implication that, regardless of whether those concerned go overseas, or not, Australian universities are wasting talents by enforcing a retiring age of 65. We hear much of the brain drain; here is another channel to it.

The editor went on to decry a system where "research scientists in full possession of their powers should be arbitrarily cut short" at 65 for men and 60 for women. According to the national viewpoint Australia could little afford to do without the skills and knowledge of older Australians. He called for a more flexible approach to retirement. (One that was finally introduced in the 1990's). Retirement at age 65 was something that historically belonged to the nineteenth century. The best way put forward was for each person to decide when they want to retire and begin taking the pension – some may want this earlier than others. "The Federal Government should appoint a commission to examine the whole question, reflecting as it does so that politicians, should share with the rest of the community their privilege of not being automatically discarded because of age." [71]

Peter Bishop succeeded him as head of the Department of Physiology. "Whereas Eccles had studied motor neurons in the spinal cord to probe the workings of the brain, Bishop and his colleagues studied sensory cells in the visual system."[72]

Before he left Canberra, Judy Cassab, a famous Australian portrait artist, was commissioned by the ANU to paint Professor Eccles. She commented in her daily diary:

> I paint Sir John Eccles for the Australian National University. I know he won the Nobel Prize, but for what? Last night he greeted me stiffly in a dark suit. In his lab this morning, in his own environment, his eyes spark fire at the adventures of every day. He is sixty-three years old and six feet tall. He moves so fast, his white smock flies. His assistants are from Tokyo and Pisa. Retiring from the ANU, he will be welcome in Chicago.
>
> Would I like to view one of the experiments after the sitting? Yes.
>
> An anaesthetised cat lies on the table, head open, brain showing electrodes connected to nerve centres and as they receive the electric stimulus, a closed-circuit television screen shows the secret signs as each zigzag represents a message of the animal's brain.[73]

---

[70] The Sun, 17 February, 1966.
[71] The Canberra Times, 17 February 1966.
[72] The Making of an Australian University, 1996:381.
[73] Judy Cassab Diaries, 1995:176.

Photograph by courtesy of John Curtin School of Medical Research.

At his farewell dinner, Jack Eccles is presented with a large cartoon by Frith depicting his taking off to America with two cats in his luggage and another peeking out of his head. Others in the photo, from left to right, are Judy Cassab, the artist who painted his portrait, Professors Hugh Ennor, Leonard Huxley and Frank Gibson.

From 1966 to1968, Jack Eccles was a Member of the Institute for Biomedical Research in Chicago, USA. In his *Scientific Odyssey*, Eccles described his research time in Chicago as being brief, unsuccessful and unhappy and he failed to gain an extension of the retiring age of 68, which prevailed there. He then moved to the State University of New York at Buffalo where the research facilities were on a par with the John Curtin School of Medical Research in Canberra. During his years in America, he had 20 collaborators from 11 countries, including Japan, Sweden, Australia. "Their cerebellar research --- was an important component of the very large amount of new and detailed information, including the discovery of several governing principles and cellular mechanisms, that Eccles contributed to the understanding of the cerebellum and its associated structures." [74]

It was during his time in Chicago that Sir John was divorced from his wife, Rene, after nearly forty years of marriage when the youngest of their nine children, Richard, was twenty-one years old. In 1968, Jack married Dr Helena Táboriková who was a member of his research team.

In 1969, Jack Eccles was invited to give the Foerster Lecture at Berkeley with the title *The Brain and the Soul*. His lecture drew a large audience and had to be held in a lecture room large enough for 3000. Until then, Sir John had been reluctant to use the word, soul, instead using the term self. In his lecture, he carefully established his academic credentials as a brain scientist before introducing the notion of soul:

> I believe that my experiencing self (soul) is only in part explained by the evolutionary origin of my body and brain, --- The uniqueness that I experience cannot be attributed to the uniqueness of my genetic inheritance, as I have already argued. Our coming-to-be is as mysterious as our ceasing-to-be at death.

In 1972, Sir John wrote to his daughter, Mary, from the Laboratory of Neurobiology, State University of New York. "How very good of you to send me the genealogical tree. I am not sure what its implications are. It always amazes me that at the end of all these vicissitudes, even in historical times, we actually come to exist." At this stage he had two books on the way. One being printed was *The Self and Its*

---

[74] Historical Records of Ausralian Science, Vol13, No. 4.

Brain with Karl Popper and another one he was about to begin was the biography of Sherrington with Bill Gibson. "We work at the magnificent Villa Serbelloni on Lake Como."

His volume of work continued unabated, as he noted in the same letter: "Here we carry on with research still – perhaps for three more years before retiring, but life is extremely busy and fruitful. Some two years ago I published a book, *Facing Reality* and now have another one, *The Understanding of the Brain,* in preparation. But of course I have much research to write up and many lectures to give. Last year (1971) I gave 61 lectures in one or another place in Europe and the U.S.A."

Facing Reality was first published in 1970. The word 'facing' in the title is said to be less defiant than the word 'confronting' because, as Eccles put it, " [In] life with its joys and sorrows, its successes and failures, its peace and its turmoil, my attitude is one of serene acceptance and gratitude and not one of angry confrontation and rejection. Reality [is] what we each of us must face if we are to live and adventure as free and responsible beings".[75]

In 1973, *The Understanding of the Brain* was published while Sir John was working in Buffalo and was based on the thirty-third Series of Lectures of the Patten Foundation delivered from 22 February to 4 April 1972. In this book, he tried to convey to the audience that, "the scientific study of the brain is an exciting experience, and that we are only at the beginning of this, the most wonderful adventure that man can undertake, since its aim is to understand man. As I have questioned above: How far will it be possible for us using our brains to understand our brains?"

His first Gifford Lecture series was published in a volume called *The Human Mystery* in 1979 and further Gifford lectures were published in 1980 as *The Human Psyche,* which examines materialist theories. Eccles tried to show that dualist-interactionism offers valuable insights into the higher levels of human experience that cannot be accommodated by materialist theories of the mind. Lord Gifford had founded these lectures for the purpose of having Natural Theology treated as a science just as Astronomy or Chemistry. Eccles noted that, in his lecture series, he would endeavour to create, "an atmosphere of wonder and of humility before the grandeur and immensity of the great cosmos". [76]

The Self and its Brain, written with Sir Karl Popper, was published in 1977. According to the jacket, "this book creates the first link between the philosophy of self and neurobiology. In dealing with the self, philosophers of the self and neurobiology have so far taken little account of scientific knowledge of the brain; scientists have traditionally avoided philosophy in favour of purely material evidence. Eccles, a neurobiologist and Popper, a philosopher both believers in dualism and interactionism consider the existence of consciousness one of the greatest riddles of cosmology." [77]

When *Sherrington, his Life and Thought* appeared, Eccles was, at last, able to return the gratitude he had felt all his life towards this great man. It was written in conjunction with Bill Gibson who had shared Eccles' time in Oxford.

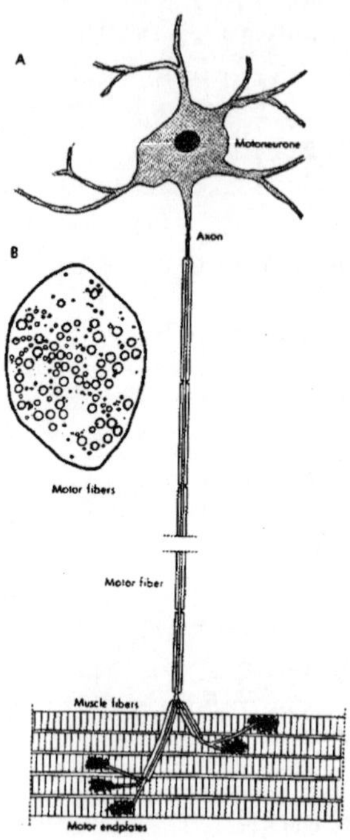

The motor unit. A: Motoneurone with its axon passing as a myelinated nerve fibre to innervate muscle fibres; B: Transverse section of motor fibres supplying a cat muscle, all afferent fibres having degenerated.

(J. C. Eccles and C. S. Sherrington. Proc. Roy. Soc, B 106:326 (1930). Quoted in the "Understanding of the Brain", J. C. Eccles, 1973:35)

---

[75] Facing Reality, 1973.
[76] The Self and its Brain, 1977.
[77] The Self and its Brian, 1977

Chapter 7:

Switzerland, 1975 to 1997:

In 1975, Sir John Eccles and his wife, Helena, retired to a lovely home outside Contra Village, near Locarno, in Switzerland. The house was flat roofed which meant that in heavy snow storms, the roof had to be cleared of the weight of snow, but it was a beautiful location with wild flowers everywhere and roses growing over fences in the summer. Behind the house was the usual vegetable garden, orchard and swimming pool. A table under a grapevine was a favourite place to have meals or to write papers in the summer heat, overlooking the lake and with the snow-topped mountains beyond.

John Eccles in a Swiss village, 1980, when the author visited him at his home.

Sir John, often accompanied by Lady Helena, still travelled overseas giving lectures and attending conferences as can be seen by snippets from letters to his children.

In 1979 to Mary:

> I have just handed in another book, *The Human Psyche,* to follow the first Gifford Lecture *The Human Mystery.* It was quite a big struggle.

Again in 1982,

> On Sunday we go to Houston Texas for a month of hard work in the big institute there working on motor disabilities of all kinds. Last year (81) was too hectic – away for 26 weeks at conferences and meetings etc. This year will also be busy.

In 1981, Sandy Rovner interviewed Eccles for the *News Weekly* at the Georgetown University in Washington where he had come to deliver a series of lectures on *The Nature of Brain and the Problem of Mind.* She headed her article *Sir John Eccles, the Evolutionist who says evolution is not the final story*:

> Sir John won the Nobel Prize for medicine in 1963 for his work in demonstrating the transmission of electrical impulses in the brain. He probably knows more about how the brain works than anyone, about its 10 billion (or so) elements organised into 3 million (or so) modules. So when he says that his understanding of the brain and how it evolved does not begin to explain the genesis of human consciousness of self, one is compelled to listen. Sir John has become a philosopher, a metaphysician, drawing on the best thought of Western civilisation to lead him, with philosopher Karl Popper, to the concept he calls dualist-interactionism. His work of the past decade or so has led him to conclude that evolution alone cannot explain human awareness of self, that there must have been the intervention of some transcendental agency, of God, in the infusion into humans of soul. [78]

---

[78] News Weekly, 29 April 1981.

In 1985, he returned to Georgetown University to receive an Honorary Doctorate, which was bestowed as recognition for "a lifetime of service to truth and service to what is most ennobling in our species".

The citation read in part:

> Sir John stands today as one of the premier scientists of our age and surely one of its most devoted and inspiring humanists. Over a period of more than half a century he has enlarged our understanding of the principles and processes governing the activity of the brain. To these scientific facts, unearthed through extraordinary research, he has added what is perhaps an even rarer contribution: a deeper appreciation and respect for the Wonder of Life, for the preciousness of the gift of life and especially the gift of mind.[79]

His eldest son, Dr Peter Eccles, a physicist, who was working at the Massachusetts Institute of Technology, was able to be present at the ceremony. Before this Peter had called his father, "to ask if he would like to be a MITRE Distinguished Lecturer. He responded positively, figured out a title for his talk to a bunch of engineers and computer experts, *The Neural Machinery of the Brain*. His talk was a great success". Peter noted "His talks, and debates with the other panel members filled the Physics Department Auditorium, the largest in the University, to overflowing. There was tremendous interest in his message, and he was very generous in the time he spent answering questions from the floor".[80]

The hectic pace of Sir John's life in the 1980's continued. On 10 September 1983 he wrote to his daughter, Frances, who worked as a librarian and who had a great interest in birds and native plants:

> Here I am in surprisingly good health and have much work to do both physical in the large garden and mental in the many writings and lectures that I give – including a new book with Professor Dan Robinson of Georgetown University, Washington, D.C. It is entitled, *The Wonder of Being Human*, and is designed for a large audience that keeps on asking the Great Questions and is searching for answers that often are phony or science fiction or pure materialism. The book comes out in the USA and England late this year and in Germany and Italy next year in translation. The book with Karl Popper sells surprisingly well in the German translation – about 30,000 copies in 10 months in hard cover. I tell you all these details because you are a librarian!

Again to Frances on 31 December 1983:

> I write at the end of this year – my 80th birthday year that was filled with much entertainment and work. I am happy to say that my constitution has stood up to it magnificently. I have still to work very hard about 80 hours a week to keep up with all my writing.

Sir John met with some of his children at various times. He had regular contact with Peter when they both lived in America. In May 1989, his daughter Alice and her husband, Kevin Ley, met up with him and Helena on a visit to Locarno. They spent a pleasant evening walking the streets of the city. It was the Eve of the Ascension and the local people were crowding in anticipation of the feast. Rose and Richard met him in England and Mary visited him in Switzerland on three occasions.

In 1993, his 90th birthday was celebrated with his wife Helena and the local dignitaries in Ticino, Switzerland where he was made an honorary citizen of the Tenero-Contra commune and where he was regarded as a grand old man. There was snow on the ground and the houses as it was 27 January 1993 and a cold winter's day. About the same time, the occasion was celebrated at Melbourne University on a hot summer's day when the inaugural Eccles Lecture in Neuroscience, entitled *Stabilising Reflexes in Humans*, was delivered by Professor Ian McCloskey, the Director of the Prince of Wales Medical Research Institute, Sydney. The lecture was part of the Australian Neuroscience Society meeting. Forty-three of Sir John's descendants and relations came to celebrate the occasion with the scientists.

---

[79] Citation, Honorary Doctorate, Georgetown University, 1985.
[80] Personal communication, Dr Peter Eccles.

The five daughters of Sir John Eccles with Dr John McCloskey who presented the first Eccles Lecture at Melbourne University, 1993. From left, Alice Ley, Mary Mennis, Dr John McCloskey, Frances Standing, Judith Ryan and Rosamond Mason.

It was noted that Sir John and Rene's children have all done well in their various chosen fields of endeavour. Between the nine of them, there are: two Ph.D graduates, Rosamond and Peter; one veterinary surgeon, John; two librarians, Alice and Frances; one successful businessman, William; one nursing sister, Judith; and two teachers, Mary and Richard. At that time, there were thirty grandchildren and five great grandchildren.

Previously, in October 1988, Sir John wrote to his youngest son, Richard, a well-known science teacher in Canberra:

> I am in surprisingly good health and able to work 80 hours a week. The last two years I have been deeply involved in a new book *Evolution of the Brain: Creation of the Conscious Self* - I am a Darwinist for the first part but strongly advocate the Divine Creation of the soul or self in the second part. I had difficulty in getting it published. The materialists realize that I am their greatest enemy but I struggle on and it was accepted by Routledge, London in January. It is going well in the press, but there is a terrible lot to do in all the details. It probably is my last book. A big book that I published last year on the *Neurobiology of the Brain* with my good friends the McGeers, has already sold more than 7000 copies. It is certainly the best book in the world on this basic field.

Again, in May 1990, he wrote to Richard of the progress with his new book:

> Here my life is very full. My new book, *Evolution of the Brain: Creation of the Self,* is now out in two languages – four more: French, Italian, Spanish, and Japanese to come. It was quite well reviewed by a materialist, John Young, in Nature, 8 March. Now that I have been appointed as the First Address lecturer at the Farmington Institute for Christian Studies at Oxford, I lecture from 23 to 29 November (1990) at Manchester College. My life long struggle has been against materialism and for dualism and the reality of the spirit and I get quite a reputation for this.

In September 1991, Sir John and Christoph von Campenhausen participated in a two-week summer university course to promote gifted students in German Universities. Since most of the students understood English he was able to give a lecture on his latest book *Evolution of the Brain: Creation of the Self.* Afterwards he met Friedrich Beck, a quantum physicist and head of the Department of Theoretical Nuclear Physics in Darmstadt, who "completely understood the difficulties preventing further progress in the quantum physics of the mind-brain problem". Eccles was excited. He saw it as "a romantic triumph after my long series of rejections".[81]

In 1994, at the age of ninety-one, he published his last book entitled, *How the Self Controls Its Brain.* In this book, he mentions a work by another distinguished quantum physicist, Henry Margenau, *The Miracle of Existence.* It was a light at the end of the tunnel for him because here was someone who backed up his theories.

Margenau wrote:

> The mind may be regarded as a field in the accepted physical sense of the term, but it is a non-material field, its closest analogue is perhaps a probability field ... nor is it required to contain energy in order to account for all the known phenomena in which mind interacts with the brain. [82]

Earlier Jack had stated his own Philosophy:

> As a dualist I believe in the reality of the world of mind or spirit as well as in the reality of the material world. Furthermore I am a finalist in the sense of believing that there is some design in the processes of biological evolution that has eventually led to us self-conscious beings with our unique individuality and we are able to contemplate and we can attempt to understand the grandeur and wonder of nature. My aim is to review the sense of wonder and of mystery in our human existence. We must not claim to be self-sufficient --- we are creatures with some supernatural meaning as yet ill defined. .... Each of us can have the belief that we are acting in some great unimaginable supernatural drama. We should give all we can in order to play our part. [83]

In looking ahead to the time of his own death, he wrote:

> This whole cosmos is not just running on and running down for no meaning. In the context of natural theology we come to the belief that we are creatures with some supernatural meaning that is as yet ill defined. We cannot think more than that we are all part of some great design. Each of us can have the belief of acting in some unimaginable supernatural drama. We should give all we can in order to play our part in this life on earth. Then we wait with serenity and joy for the future revelations of whatever is in store after bodily death. [84]

---

[81] How the Self Controls Its Brain, 1994:145-146.
[82] Ibid,1994:23
[83] The Human Mystery, 1979:8-10.
[84] The Wonder of Being Human, 1984:179.

Chapter 8:

Death of Sir John Eccles, 2 May, 1997:

In the months before he died, he was cared for in the Carita Canton Hospital, in Locarno. It was springtime and Locarno was beautiful in the spring. I (the author) visited him bringing daffodils, and yellow and white daisies, trying to bring some of the beauty of spring into his hospital room. The green leaves in the flowers were fresh gumtips grown locally and he delighted in rubbing them, smelling the fresh gum-leaf smell, "Oh Australia, Australia" he said nostalgically. Although he had left Australia in 1967, never to return, he never forgot his own country and wanted to know how everyone was including Mark Oliphant. I reminded him of his last visit to Sherrington who had said "All that is left Jack is the soul". He nodded his head in agreement and commented "Yes, that's right".

He was now 94 years old and had managed to work almost up to the end, publishing his last work when he was ninety-one years old. He died on 2 May 1997 and was buried in a small cemetery near the church in Contra, Switzerland.

A Memorial Mass was held for Sir John at St Christopher's Cathedral, Canberra, shortly after his death. Members of his family, colleagues and friends attended. The celebrants were Fr Ken Heffernan and Fr Ian Howells SJ, who was a nephew of Sir John's. The eulogy was given by Dr David Curtis, a long time friend and colleague. Following the Mass, the family gathered at the Sir John Eccles Medical Sciences Library, which was renamed in his honour.

Photograph of Sir John Eccles' grave taken by the author when she visited Contra in August 2002. The grave overlooks Lake Maggiore and the snow-capped mountains beyond.

Obituaries:

Obituaries appeared in all the main papers in England, America and Australia.

The Australian:

> Sir John Eccles stood tall in the front rank of Australian scientists. In a brilliant career of demanding experimental work spanning the decades from 1929 to 1972, he became a giant among the leading international figures in brain research. [85]

The Medical Observer, quoting Sydney University Professor of Neurophysiology, Max Bennett:

> Eccles was remarkable for his experimental drive, achieving fame for his capacity to work in 24 hour shifts. ... His drive alone was infectious. He had tremendous charisma in terms of exciting people's interest in the brain. All of this came together in the fact that he was a Catholic. Although he divorced and remarried in the latter part of his life, his Catholicism had a tremendous effect on him. He held the dualistic position of Descartes, that there is a separation between the mind and the brain in terms of their material. [86]

The Times:

> By the end of his long life, John Eccles had become an endangered species, a scientist who believed in the distinction between the mind and brain. Though still the common belief of many people, this dualist view had fallen out of favour among physiologists: Eccles was amongst its last defenders publishing a series of books that rejected the materialist view of the mind and attempted to redefine consciousness "in terms of quantum theory". Most of his peers believed this to be simply wrong, but Eccles was not discouraged. ....
>
> Eccles' belief that the mind can operate independently of the physical processes of the brain was shared by Popper, and they collaborated in *The Self and its Brain,* published in 1977. With the German physicist Friedrich Beck, Eccles developed a physical theory to explain how this might happen. Eccles claimed that the nerve cells fire when ions accumulate at a synapse; but the mere presence of a given number of ions is not enough to trigger the firing. This was because, in his view, the ions' existed in a state of quantum uncertainty in which firing was not inevitable, but was in fact determined by the mind consciously deciding which neurons should fire and which should not.
>
> He carefully admitted there was no proof of this but claimed that it would lead to a revival of the dualist hypothesis. His motivation was religious; he refused to acknowledge that states of mind and belief were simply a reflection of the physical conditioning of the brain, which seemed to him to deny free will.[87]

The Guardian:

> His influence in science and on Australian academic structures was crucial during the 1950's: through contributions to understanding of the biochemistry of the creation of nerve signals and through the founding of the Australian Academy of Science (modelled on the Royal Society) and his vigorous aspirations for international stature in the sciences at the Australian National University. Sir John, bespectacled and firm-jawed, was a tough, determined realist, a brilliant if obstinate scientist and a shrewd politician. -- Sir John was an uncomfortable giant who never lost his rough edges; he created new and important science and helped bring valuable maturity to academic thought in Australia. [88]

---

[85] The Australian, Tuesday 6 May 1997.
[86] Medical Observer, 11 July 1997:91.
[87] The Times, May 1997.
[88] The Guardian, The Nerve of the Man, 7 May 1997.

Photograph by courtesy of John Curtin School of Medical Research.

In May 1997 at the opening of the Sir John Eccles Medical Sciences Library, John Curtin School of Medical Research, Sir John's children with the 1967 portrait by Judy Cassab.

Left to right: Dr Rosamond Mason, Mary Mennis, Richard Eccles, Alice Ley, Judith Ryan, Dr Peter Eccles and Frances Standing. Inset: William Eccles and Dr John Eccles.

The following year, 1998, the Neuroscience Laboratory was named after him in a ceremony attended by many scientists and politicians.

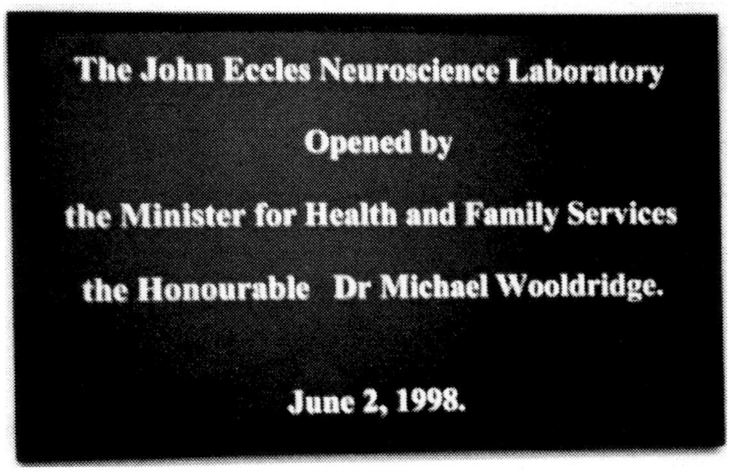

Photograph by Damian White, by courtesy of the Warrnambool Standard

Koroit School children with plaque for the former Eccles house, November 2002. Koroit Primary School Pupils, Tim Kennedy, 11, Shelli King, 11, and David Kennedy, 9 talk with Dr Arthur Christopoulos about the school's famous alumnus, Nobel Prize winner Sir John Eccles.

The inscription on the plaque reads: "The former home of Sir John Eccles 1903 – 1997. From Koroit schoolboy to Nobel Prize winning neuroscientist in Physiology and Medicine, 1963. Koroit and District Primary School - Anthony Dowling Principal. Australian Institute of Political Science - John Best Chairman. The Tall Poppy was created by the Australian Institute of Political Science as a metaphor for excellence and endeavour and symbolises Australia's pride in its outstanding achievers in all fields".

Dr Rob Wallis unveils the plaque for the Eccles Science Centre at Warrnambool College, where Sir John was a High School student from 1915 to 1918.

Epilogue, 2003:

Centenary Celebrations of Sir John's birth:

While this Portrait of Sir John Eccles has been prepared for release before the centenary of his birth on 27 January 2003, several commemorative events have already taken place.

Warrnambool College:

One of the occasions was the opening of the Sir John Eccles Science Centre at the Warrnambool College where John Eccles was a student from 1915 to 1918. It was opened by Professor Rob Wallis. a scientist from Deakin University. Richard Eccles, youngest son of Sir John, and a science teacher at the St. Francis Xavier College in Canberra, gave the response, praising the facilities of the science block and stressing the importance of science. At this gathering were former colleagues of Sir John, the Mayor of Warrnambool and other City Council representatives, members of the School Council and student leaders as well as members of the Warrnambool College Foundation.

That same evening, the inaugural Eccles Tall Poppy awards were given to seven outstanding past students of Warrnambool College. The recipients were: Astrophysicist Alicia Oshleck, who has worked at many observatories and who is completing her PhD at Melbourne University; Dr Michael Burnet, who studied biochemistry at the University of Adelaide and who now works in Germany; Finance analyst Leigh Cronin, who heads a team of analysts in Australia, New York, London and Singapore; Dr Julianne Lynch, lecturer in Education Studies at Deakin University; Dr Selby King, who is a medical practitioner in Warrnambool and also a well known conservationist; Hearing Researcher Maria Remine who is investigating the role of intelligence and higher order cognitive functions in the language abilities of hearing impaired children; and Dr Tiffany Walsh, who gained a PhD from Cambridge University and lectures in the University of Warwick in England.[89]

The Tall Poppy Symbol arose from the Florey Centenary Celebrations in 1998. It is fostered by the Australian Institute of Political Science and aims to create an environment and culture in Australia that recognises, values and supports achievement in science, so that intellectual achievement is as valued by Australians as achievement in sport and artistic endeavours.

Warrnambool College Assistant Principal, Mary Fielding, has stated that these awards would become a regular biennial event to encourage children to strive for better results in their studies.

Warrnambool primary schools:

Primary schools in Warrnambool area were invited to participate in a Tall Poppy inter-school science competition. Three schools in the District, with seven teams between them, took part. The schools were Warrnambool Primary, Warrnambool East Primary and Caramut Primary. Warrnambool Primary School's teams gained the first three places.

Koroit Primary School:

With great gusto, the Koroit Primary School celebrated Sir John Eccles Day, named after the Koroit student who went on to receive the Nobel Prize for Medicine and is considered a Great Australian Tall Poppy. John Eccles lived in the head teacher's house next to the school (see Chapter 1) where his father, William Eccles, was headmaster from 1911 to 1918 and where he was a pupil from 1911 to 1914. A special plaque, giving information about him was unveiled by Dr Arthur Christopoulos, who then addressed the children and answered their very searching questions about science. Children dressed up in traditional clothes and studied the past history of the school.

---

[89] Warrnambool Standard, 14 December 2002.

John Curtin School of Medical Research:

The **Eccles Centenary Symposium** is scheduled for 27th January 2003 in the Florey Theatre at the John Curtin School of Medical Research. The Vice Chancellor of the Australian National University, Professor Ian Chubb, will give the opening address. Presentations will be made by scientists who worked with Sir John Eccles in Canberra and in the USA. They are now all world leaders in Neuroscience and it is a tribute to Sir John that they have come to celebrate his centenary.

Among them are:

Professor Per Andersen, Dept of Neurobiology, University of Oslo, Norway;

Professor Masao Ito, RIKEN Frontier Research Program, Wako, Japan;

Professor Piergiorgio Strata, University of Turin, Italy;

Professor William D. Willis, Dept. Anatomy and Neuroscience, University of Texas Medical Branch, Galveston, USA;

Professor Robert Schmidt, Physiologisches Institute, Wurzburg;

Professor Donald Faber, Dept. of Neurobiology and Anatomy, Allegheny University Health Science, New York, USA;

Professor Mark Rowe, Department of Physiology & Pharmacology, University of NSW;

Professor David Curtis, Neuroscience Division, John Curtin School of Medical Research, Canberra, Australia.

In addition, current researchers will present papers on their work which is a direct lineage from Sir John's work. Many of these researchers work at the Neuroscience Division at the John Curtin School of Medical Research, including Dr Greg Stuart; Dr John Bekkers; Professor Bruce Walmsley; Professor Stephen Redman; Dr Pankaj Sah and Dr John Clements. In addition, Professor Peter Gage, Division of Molecular Bioscience, John Curtin School of Medical Research, and Professor Max Bennett of Department of Physiology, University of Sydney will also present papers.

Following these presentations, Professor Sam Berkovic, Eccles Lecturer for 2002 from the Epilepsy Research Institute, Austin Hospital, Melbourne, will speak on Single Gene Epilepsies.

On that same day, members of the Eccles family and the participants in the Symposium will attend the unveiling of the plaque marking the Centenary. Later in the evening, Tall Poppy Awards will be presented to five outstanding achievers.

Appendix:

The following are some of the Honours and Honorary Memberships received by Sir John Eccles during his lifetime.

Honours:

1931 Rolleston Prize.

1941 Fellow of the Royal Society.

1941 Royal Medal.

1958 Knight Bachelor.

1963 Nobel Prize for Physiology or Medicine.

1963 Australian of the Year.

1963 Mitglied Deutsche Academie Naturforschung and Cothenius Medal.

1973 Foreign Member Max-Planck-Inst. d. Biophysik, Chemie, Göttingen.

1986 Order of the Rising Sun (Gold and Silver Stars) Japan.

1990 Companion in the Order of Australia.

1991 Cortina-Ulisse Literary Prize.

Honorary Memberships:

Some of his Honorary Memberships of International Scientific Societies were:

National Academy of Science.

American Philosophical Society.

Academia dei Lincei.

Royal Belgian Academy.

Bavarian Academy.

American Academy of Art and Sciences.

American College of Physicians.

Pontifical Academy.

Indian Academy of Science.

Academia Europea.

Honorary Doctorates:

He received Honorary Doctorates from many prestigious Universities, including:

Cambridge	Melbourne	Tasmania
British Columbia	Marquette	Loyola
Gustaphus Adolphus	Yeshiva	Prague
Oxford	Fribourg	Torino
Georgetown (Washington)	Tsukuba (Japan)	Basel

Bibliography:

The following is a selected list of sources consulted in the compilation of this biography.

Books:

Cassab, Judy, *Diaries*, Australia Council for the Arts, 1995.

Curtis, D. and Per Andersen. *Historical Records of Australian Science*: Australian Academy of Science, Canberra: 2001. Vol 13 Number 4.

Eccles, J.C. *Brain and the Conscious Experience.* A Study week September 28 to October 4, 1964 of the Pontificia Academia Scientiarum: Springer-Verlag. Berlin. 1966.

Eccles J. C. *The Neurophysiological Basis of Mind; The Principles of Neurophysiology. The Waynflete Lectures, 1952*, Oxford, Clarenden Press, 1953.

Eccles J. C. *The Physiology of Nerve Cells*:. Baltimore John Hopkins Press 1968.

Eccles J. C. *Facing Reality*: Roche Basel 1973.

Eccles J.C. *My Scientific Odyssey*. A. Rev. Physiology: 1977, 39, 1-18.

Eccles J.C. & Karl Popper *The Self and Its Brain*: Springer International 1977.

Eccles J.C. *The Human Mystery.* Springer International. 1979

Eccles, J.C. *The Wonder of Being Human. Our Brain and Our Mind.* Free Press. London.1984.

Eccles J.C. *Evolution of the Brain: Creation of the Self*: Routledge, London. 1989.

Eccles J.C. *How the Self Controls the Brain:* Springer-Verlag 1994.

Encyclopaedia of World Biography Vol 5.

Forster S.G. & Varghese M: *The Making of the Australian National University*: Allen and Unwin 1996.

Heritage 200 Committee. *People Who Made Australia Great.* Collins Australia 1988.

Home R.W. (Editor), *Australian Science in the Making*, Cambridge University Press, 1988.

McCorkell and Peter Yule. *A Green and Pleasant Land: A History of Koroit.* 1999.

Mennis M. *Hullo Eccles. The Eccles family 1850 – 2000*: Lalong Enterprises 2000.

Murphy John. *No Parallel. The Woorayl Shire 1888-1988.* Hargrave Publishing Company 1988.

Sherrington C. *Man on his Nature*: Cambridge University Press, Cambridge 1940.

Shire of Korumburra. *The Land of the Lyrebird. A Story of Early Settlement in the Great Forest of South Gippsland*: Published by the Shire of Korumburra for the South Gippsland Development League. *1920.*

The Old Unicornian, the magazine *of* Melbourne High School.

The Speculum., the magazine of the Melbourne Medical Students' Society.

Victorian Education Department, *Vision and Realization. A Centenary History of State Education in Victoria.* Vol 3. Education Department of Victoria 1973.

Newspapers and Journals:

Journal of Neurophysiology; Medical Observer; Newsweekly; The Advertiser; Nature

The Guardian, Sydney Morning Herald; The Australian; The Canberra Times,

The Sun, Melbourne, Sun Herald; The Warrnambool Standard.